MANUAL OF PLANKTONIC FORAMINIFERA

MANUAL

OF

PLANKTONIC FORAMINIFERA

BY

J. A. POSTUMA

Royal Dutch/Shell Group, The Hague, The Netherlands

ELSEVIER PUBLISHING COMPANY

AMSTERDAM, LONDON, NEW YORK 1971

ELSEVIER PUBLISHING COMPANY
335 JAN VAN GALENSTRAAT
P.O. BOX 211, AMSTERDAM, THE NETHERLANDS

ELSEVIER PUBLISHING CO. LTD.
BARKING, ESSEX, ENGLAND

AMERICAN ELSEVIER PUBLISHING COMPANY, INC.
52 VANDERBILT AVENUE
NEW YORK, NEW YORK 10017

LIBRARY OF CONGRESS CARD NUMBER: 75-151739

ISBN 0-444-40909-2

WITH 1153 ILLUSTRATIONS

PRINTED IN THE NETHERLANDS

PREFACE

The use of planktonic Foraminifera as guide fossils is generally accepted today. Their abundance in open marine environments, and the short stratigraphic ranges of many species, make this group of Foraminifera one of the most suitable and reliable tools for detailed biostratigraphic investigations. The planktonic Foraminifera, of practical use in biostratigraphy, first occur during the Early Cretaceous. They continue on a world-wide scale, and in a rapid succession of species, to the Recent.

During the last twenty years an ever-increasing flow of publications has appeared on this subject. Among the great number of new species and subspecies described, many proved to be useful guide fossils, others turned out to be synonyms and many cannot be recognized by other workers in this field. A number of varying classifications on the supra-specific level have also been proposed in recent years. This demonstrates that there still is disagreement on the genetic and phylogenetic relations of the planktonic Foraminifera. Furthermore, authors in general do not hold the same views on the definitions of species. Such diverse treatment of this stratigraphically important group of Foraminifera is bound to lead to considerable confusion. As a result, it has become almost impossible for the industrial palaeontologist to judge all published data on their merits, a task which, in addition, is not rendered any easier by their being distributed over a large and scattered number of publications. In 1960, it was decided to compile a manual that would provide a selection of clearly defined species to enable the operational palaeontologists of the Royal Dutch/Shell Group to make effective use of planktonic Foraminifera in obtaining more precise and uniform stratigraphic results. A growing interest in this manual outside the Group induced the Bataafse Internationale Petroleum Maatschappij N.V. to give permission for the publication of this work.

It should be mentioned that no particular published classification of genera has been followed here. Many classifications have been published and in view of the new vistas opened by the scanning electron microscope we may certainly expect the introduction of others in the near future. However, the principles put forward in the outstanding publications of BOLLI, LOEBLICH and TAPPAN (1957) and BANNER and BLOW (1959) have been largely adopted. The use of subspecific names has been abandoned in order to improve communication between palaeontologist and geologist. In principle, the idea behind this work was to enable the user to arrive at a quick and reliable age deter-mination of his planktonic faunas. Therefore, the complex family Heterohelicidae, for example, is not included in this manual, nor are species of uncertain stratigraphic range or those which are difficult to recognize.

Special attention has been paid to the illustrations of the species. Planktonic Foraminifera are often abundant in carbonate rocks, where they can only be studied in thin section. For this reason, a figure of an axial section of each species has been provided. In addition, thin sections of hard rocks showing typical planktonic assemblages have been included where it is thought they would be useful.

It should be realized that the distribution of planktonic Foraminifera is controlled by several factors, one of the most important of which is the climatic influence. The foraminiferal zonation, consisting of 48 zones, proposed in this manual is only valid for tropical and sub-tropical environments. This restriction is most strongly manifested in the Upper Miocene–Recent interval, during which time the tropical and sub-tropical belt contracted.

The writer thanks the Management of the Bataafse Internationale Petroleum Maatschappij N.V. (Royal Dutch/Shell Group), for permission to publish this manual. He is indebted to Dr. C. W. Wagner, the originator of the project. Many thanks are due to Prof. Dr. H. M. Bolli, who has given valuable advice and assistance with the preparation of the original B.I.P.M. manual between the years 1960–1966. The writer is particularly grateful to Mr. W. Geluk, who made all the photographs and to Mr. M. van Dugteren whose excellent drawings are such an important contribution to this work. Mrs. Th. M. Jansen and Mrs. A. M. Anjema should be mentioned for their skill and patience during the preparation of thin sections of isolated planktonic specimens. Finally the writer wishes to express his thanks to Miss W. Mostert who typed the manuscript.

The Hague, March 1971

CONTENTS

I. MESOZOIC
(Albian–Maastrichtian)

KEY TO THE MESOZOIC GENERA

I Test trochospiral

    A Primary aperture umbilical, with tegilla
        1 with keel(s)   — *Globotruncana* CUSHMAN, 1927
        2 without keel(s) — *Rugoglobigerina* BRONNIMANN, 1952

    B Primary aperture extraumbilical-umbilical, with sutural supplementary apertures
        1 with keel   — *Rotalipora* BROTZEN, 1942
        2 without keel   — *Ticinella* REICHEL, 1950

    C Primary aperture extraumbilical-umbilical, bordered by a narrow lip or spatulate flap
        1 with keel   — *Praeglobotruncana* BERMUDEZ, 1952
        2 without keel   (a. chambers globular to ovate —
                                Hedbergella BRONNIMANN and BROWN, 1958
                              b. later chambers clavate to radial-elongate —
                                  Clavihedbergella BANNER and BLOW, 1959

II Test in early portion trochospiral, later stage planispiral
    Aperture extraumbilical, tending to become equatorial

        1 chambers elongated, with a hollow bulb-shaped or spine-like extension in the equatorial plane — *Schackoina* THALMANN, 1932

        2 chambers elongated; some or all chambers of the last whorl with two, or occasionally more, hollow bulb-shaped extensions on each side of the equatorial plane — *Leupoldina* BOLLI, 1957

III Test planispiral

    Primary aperture equatorial bordered by a lip, with relict apertures
        1 with keel   — *Planomalina* LOEBLICH and TAPPAN, 1946
        2 without keel   (a. chambers subglobular to ovate —
                                Globigerinelloides CUSHMAN and TEN DAM, 1948
                              b. chambers radial-elongate —
                                Hastigerinoides BRONNIMANN, 1952

DESCRIPTIONS OF THE MESOZOIC GENERA

Genus *Clavihedbergella* BANNER and BLOW, 1959

Reference BANNER, F. T. and BLOW, W. H., 1959: The classification and stratigraphical distribution of the Globigerinaceae. — Palaeontology, 2 (1):18.

Type species *Hastigerinella subcretacea* TAPPAN, 1943.

Diagnosis Test low trochospiral, biconvex, umbilicate; periphery rounded and deeply lobulate, no keel or poreless margin.
Wall finely perforate, surface smooth to hispid.
Chambers of the first whorls globular to ovate, chambers of the last whorl clavate to radial-elongate.
Sutures radial, straight to curved.
Aperture an interiomarginal, extraumbilical-umbilical arch, with a narrow bordering lip or spatulate flap.

Remarks *Clavihedbergella* has been proposed by BANNER and BLOW (1959) as a subgenus of *Praeglobotruncana*; it is here raised to generic status.

Genus *Globigerinelloides* CUSHMAN and TEN DAM, 1948

Reference CUSHMAN, J. A. and TEN DAM, A., 1948: *Globigerinelloides,* a new genus of the Globigerinidae. — Contr. Cushman Lab. Foram. Res., 24:42.

Synonymy *Biglobigerinella* LALICKER, 1948: Journ. Paleontol., 22:624.
Biticinella SIGAL, 1956: Soc. Géol. France, C. R. Somm., 3:35.

Type species *Globigerinelloides algeriana* CUSHMAN and TEN DAM, 1948.

Diagnosis Test planispiral, biumbilicate, involute to partly evolute; lobulate in outline, with no indication of keel or poreless margin.
Wall perforate, surface smooth or roughened.
Chambers rounded to ovoid, may be somewhat elongated in specimens tending to become evolute.
Sutures depressed, straight to curved or sigmoid.
Primary aperture an interiomarginal equatorial, broad, low arch bordered by a prominent lip, with lateral umbilical portions of successive apertures remaining open as supplementary relict apertures, each with a remnant of the bordering lip.

Genus *Globotruncana* CUSHMAN, 1927

Reference   CUSHMAN, J. A., 1927: An outline of a re-classification of the Foraminifera. — Contr. Cushman Lab. Foram. Res., 3 (1):91.

Synonymy   *Rosalinella* MARIE, 1941: Mém. Mus. Hist. Nat. Paris, n. ser., 12:237.
Bucherina BRONNIMANN and BROWN, 1956: Eclog. Geol. Helvetiae, 48 (2-1955):557.
Rugotruncana BRONNIMANN and BROWN, 1956: Eclog. Geol. Helvetiae, 48 (2-1955):546.
Marginotruncana HOFKER, 1956: Neues Jahrb. Geol. Paläontol., Abh., 103 (3):319.
Abathomphalus BOLLI, LOEBLICH and TAPPAN, 1957: Bull. U.S. Nat. Mus., 215:43.
Globotruncanita REISS, 1957: Contr. Cushman Found. Foram. Res., 8 (4):136.
Helvetoglobotruncana REISS, 1957: Contr. Cushman Found. Foram. Res., 8 (4):137.

Type species   *Pulvinulina arca* CUSHMAN, 1926.

Diagnosis   Test trochospiral, biconvex, spiroconvex or umbilicoconvex; wide umbilicus; periphery rounded to angular with a single keel, or truncate with a double keel; keels may be beaded.
Wall perforate, surface smooth, rugose or beaded.
Chambers ovate, hemispherical, angular-rhomboid or angular-truncate.
Sutures on both the spiral and umbilical side curved or radial, depressed or elevated; they may be limbate and beaded.
Primary aperture interiomarginal, umbilical, covered by extensions from the chambers (tegilla), which form an imperforate complex umbilical cover plate with accessory infralaminal and intralaminal apertures.

Genus *Hastigerinoides* BRONNIMANN, 1952

Reference   BRONNIMANN, P., 1952: Globigerinidae from the Upper Cretaceous (Cenomanian-Maestrichtian) of Trinidad, B.W.I. — Bulletin of American Paleontology, 34 (140):52.

Synonymy   *Eohastigerinella* MOROZOVA, 1957: Dokl. Acad. Sci. U.R.S.S., 114:1109.

Type species   *Hastigerinella alexanderi* CUSHMAN, 1931.

Diagnosis   Test planispiral, biumbilicate, stellate in appearance.
Wall perforate, surface smooth, pitted or finely hispid.
Chambers in early stage globular, later chambers radial-elongate, much produced, tapering or clavate.
Sutures radial, depressed.
Primary aperture an interiomarginal, equatorial arch, bordered above by a prominent lip; the lateral umbilical portions of successive apertures may remain open as supplementary relict apertures, each with a remnant of the bordering lip.

Genus *Hedbergella* BRONNIMANN and BROWN, 1958

Reference   BRONNIMANN, P. and BROWN, N. K., 1958: *Hedbergella,* a new name for a Cretaceous planktonic Foraminiferal genus. — Journal of the Washington Academy of Sciences, 48 (1):15-17.

Synonymy   *Hedbergina* BRONNIMANN and BROWN, 1955: Eclogae Geologicae Helvetiae, 48 (2):529.

Type species   *Anomalina lorneiana* D'ORBIGNY var. *trocoidea* GANDOLFI, 1942.

Diagnosis   Test trochospiral, biconvex, umbilicate; periphery rounded, with no indication of keel or poreless margin.
Wall finely perforate, surface smooth to hispid or rugose.
Chambers globular to ovate.
Sutures depressed, radial, straight to curved.
Aperture interiomarginal, an extraumbilical-umbilical arch bordered above by a narrow lip or spatulate flap; in forms with a broad open umbilicus the successive apertural flaps may remain visible, presenting a serrate or scalloped border around the umbilicus.

Genus *Leupoldina* BOLLI, 1957

Reference   BOLLI, H. M., 1957: The Foraminiferal Genera *Schackoina* THALMANN, emended and *Leupoldina,* n. gen. in the Cretaceous of Trinidad, B.W.I. — Eclogae Geologicae Helvetiae, 50 (2):275.

Type species   *Leupoldina protuberans* BOLLI, 1957.

Diagnosis   Test in the early stage slightly trochospiral, later planispiral.
Wall perforate, smooth, pitted or hispid.
Chambers in the early part globular to subglobular, in the last whorl elongate; some or all chambers of the last whorl have two or occasionally more tubulospines with bulb-shaped extensions (mostly broken off), symmetrically arranged on each side of the equatorial plane.
Sutures radial, depressed.
Aperture an interiomarginal, equatorial arch in the early stage; ultimate chamber with two interiomarginal, umbilical apertures, one on each side of the chamber.

Genus *Planomalina* LOEBLICH and TAPPAN, 1946

Reference LOEBLICH, A. R. and TAPPAN, H., 1946: New Washita Foraminifera. — Journal of Paleontology, 20 (3):257.

Type species *Planulina buxtorfi* GANDOLFI, 1942.

Diagnosis Test planispiral; biumbilicate, involute to partly evolute, lobulate in outline with a distinct keel.
Wall finely perforate, surface smooth.
Chambers angular-rhomboid.
Sutures curved, elevated, beaded in the first part of the last whorl.
Primary aperture an interiomarginal, equatorial arch bordered by a distinct lip, with the opening extending back at either side to the septum at the base of the chamber, the lateral umbilical portions of successive apertures remaining open as supplementary relict apertures, each with a remnant of the bordering lip.

Genus *Praeglobotruncana* BERMUDEZ, 1952

Reference BERMUDEZ, P. J., 1952: Estudio sistematico de los Foraminiferos rotaliformes. — Venezuela, Minist. Minas Hidrocarb., Bol. Geol., 2 (4):52.

Synonymy *Rotundina* SUBBOTINA, 1953: Trudy Vses. Neft. Naukno-Issledov. Geol. — Razved. Inst., n. ser., 76:165.
Globotruncanella REISS, 1957: Contr. Cushman Found. Foram. Res., 8 (4):135.

Type species *Globorotalia delrioensis* PLUMMER, 1931.

Diagnosis Test trochospiral, biconvex to spiroconvex, umbilicate; periphery rounded to subangular, with a moderate keel in the early stages, commonly progressively less prominent in the later development. This keel may be beaded.
Wall finely perforate, surface smooth to hispid or partly nodose.
Chambers ovate to subangular.
Sutures on the spiral side radial or curved, depressed to raised, sometimes thickened or even beaded, on the umbilical side depressed and radial.
Aperture interiomarginal, a relatively high and open extraumbilical-umbilical arch bordered above by a narrow lip or spatulate flap; in forms with a broad open umbilicus the successive apertural flaps may remain visible, presenting a serrate or scalloped border around the umbilicus.

Genus *Rotalipora* BROTZEN, 1942

Reference BROTZEN, F., 1942: Die Foraminiferengattung Gavelinella nov. gen. und die Systematik der Rotaliiformes. — Sveriges Geologiska Undersökning, Avh., ser. C, no. 451, 36 (8):32.

Synonymy *Thalmanninella* SIGAL, 1948: Rev. de l'Inst. Français du Pétrole et Annales des Combustibles liquides, 3 (4):101.

Type species *Rotalipora turonica* BROTZEN, 1942.

Diagnosis Test trochospiral, biconvex to planoconvex, umbilicate; periphery angular with a single keel, which is mostly beaded.
Wall perforate, smooth to ornamented with calcareous ridges or knobs both on the spiral and on the umbilical side.
Chambers angular-rhomboid.
Sutures on the spiral side curved, depressed to elevated, may be limbate and beaded, on the umbilical side radial to slightly curved, depressed to elevated, and sometimes limbate and beaded, especially in the first part of the last whorl.
Primary aperture interiomarginal, extraumbilical-umbilical, with an imperforate flap which fuses with the preceding ones; a single secondary sutural aperture per suture with a bordering lip.

Genus *Rugoglobigerina* BRONNIMANN, 1952

Reference BRONNIMANN, P., 1952: Globigerinidae from the Upper Cretaceous (Cenomanian-Maestrichtian) of Trinidad, B.W.I. — Bull. Amer. Paleontol., 34 (140):16.

Synonymy *Plummerella* BRONNIMANN, 1952: Bull. Amer. Paleontol., 34 (140):37.
Plummerita BRONNIMANN, 1952: Contr. Cushman Found. Foram. Res., 3 (4):146.
Trinitella BRONNIMANN, 1952: Bull. Amer. Paleontol., 34 (140):56.
Kuglerina BRONNIMANN and BROWN, 1956: Eclog. Geol. Helvetiae, 48 (2-1955):557.

Type species *Globigerina rugosa* PLUMMER, 1926.

Diagnosis Test trochospiral, biconvex, umbilicate; periphery rounded to angular, rarely with a pseudokeel.
Wall perforate, surface typically rugose with numerous large pustles which may coalesce into distinct ridges, radiating from the midpoint of each chamber on the periphery, or much produced peripherally into spinelike extensions, more rarely smooth.
Chambers spherical, hemispherical, radial-elongate or (rarely) angular in the later portion.
Sutures radial to slightly curved on the spiral side, radial on the umbilical side, generally depressed throughout, in the later portion of the test sometimes elevated and limbate.
Primary aperture interiomarginal, umbilical, covered by extensions from the chambers (tegilla), which form an imperforate complex umbilical cover plate with accessory infralaminal and intralaminal apertures.

Genus *Schackoina* THALMANN, 1932

Reference THALMANN, H., 1932: Die Foraminiferengattung *Hantkenina* CUSHMAN, 1924 und ihre regional-stratigraphische Verbreitung. — Eclogae Geologicae Helvetiae, 25 (2):289.

Type species *Siderolina cenomana* SCHACKO, 1896.

Diagnosis Test in the early portion trochospiral, later stage becoming nearly or completely planispiral. Wall finely perforate, smooth or finely hispid.
Chambers in the trochospiral part subglobular, in the planispiral part radially elongate, with a bulb-shaped extension (mostly broken off-as final part of a tubulospine) in the equatorial plane.
Sutures straight, radial, depressed.
Aperture an interiomarginal arch or slit, extraumbilical, tending to become equatorial; may be bordered by a lip.

Genus *Ticinella* REICHEL, 1950

Reference REICHEL, M., 1950: Observations sur les *Globotruncana* du gisement de la Breggia (Tessin). — Eclogae Geologicae Helvetiae, 42 (2):600.

Type species *Anomalina roberti* GANDOLFI, 1942

Diagnosis Test trochospiral, biconvex to planoconvex, umbilicate; periphery rounded, lacking keel or poreless margin.
Wall perforate, surface smooth or partly pustulate.
Chambers ovate to subglobular.
Sutures on spiral side curved, depressed to elevated, on umbilical side slightly curved to radial, depressed to flush.
Primary aperture interiomarginal, extraumbilical-umbilical, with an imperforate flap which fuses with the preceding ones; a single sutural aperture per suture with a bordering lip.

DESCRIPTIONS OF THE MESOZOIC ZONES
(ALBIAN–MAASTRICHTIAN)

The zonal scheme of the Middle to Late Cretaceous interval presented in this work has been compiled from local zonations published for Trinidad and North Africa, and from the Royal Dutch/Shell Group's exploration studies in such areas as Tunisia, Italy, West Pakistan and West Irian. As our knowledge of the taxonomy and the distribution of the Early Cretaceous planktonic Foraminifera is still rather fragmentary, this zonation commences in the Albian.

Chart 1 shows the zonation together with local zonations from Trinidad, north Africa and the U.S. western Gulf coastal plain and, additionally, a tentative correlation with the classic European stages and the Gulf Coast divisions. This biostratigraphic zonation, based as it is upon commonly occurring, characteristic planktonic organisms, is considered to be as near to a time-stratigraphic subdivision as is currently possible. The proposed zonation is meant to be a general framework; if local conditions allow, and if such is found desirable, the zones may be still further subdivided. A short definition of the zones is given below:

1 **Ticinella roberti** zone (Partial-range zone)
   This zone is exclusively based on the presence of *T. roberti* but with the absence of *Rotalipora*.

2 **Rotalipora subticinensis** zone (Partial-range zone)
   This zone is defined as that part of the range of *R. subticinensis* prior to the first appearance of *Globigerinelloides breggiensis* and *R. ticinensis*. *Ticinella roberti* occurs together with *R. subticinensis*, but it ranges above and below the zone.

3 **Globigerinelloides breggiensis** zone (Total-range zone)
   Rotalipora ticinensis first appears together with the zonal guide fossil at the base of the zone, but continues into the younger zones. *R. subticinensis* ranges into the lower part, while *Ticinella roberti* ranges beyond the zone. *Hedbergella washitensis* may be present in the uppermost part, while *Planomalina buxtorfi* is not yet present.

4 **Planomalina buxtorfi** zone (Partial-range zone)
   P. buxtorfi first appears at the base of the zone, but may not be restricted to it, as questionable occurrences are reported from the lower part of the *Rotalipora appenninica* zone. Such questionable occurrences in the *R. appenninica* zone are also reported for *Ticinella roberti*, which ranges all the way up from the *T. roberti* zone. *R. ticinensis* and *Hedbergella washitensis* occur in the zone but are not restricted to it.
   Globigerinelloides breggiensis becomes extinct at the lower boundary of the zone. The top of the zone is defined by the first appearance of *R. appenninica* and *Praeglobotruncana stefani*.

5 **Rotalipora appenninica** zone (Partial-range zone)
   The zone is defined as that part of the ranges of *R. appenninica* and *Praeglobotruncana stefani*, which occur before the first appearance of *R. greenhornensis*. *Hedbergella*

washitensis extends upward into this zone, but its presence in the uppermost part remains doubtful. *R. ticinensis* is known to extend into the lower part of the zone; the occurrences of *Planomalina buxtorfi* and *Ticinella roberti* in this lower part are questionable.

6 **Rotalipora greenhornensis** zone (Partial-range zone)
This zone is defined as that part of the range of *R. greenhornensis* below the first appearance of *R. cushmani*.

7 **Rotalipora cushmani** zone (Total-range zone)
In contrast to the zone's guide fossil, the even more characteristic *R. reicheli* is limited to the middle and upper parts of the zone. The latter species may therefore constitute a marker for the middle and upper part of the *R. cushmani* zone. Other species ranging into but becoming extinct within the zone are *R. appenninica* and *R. greenhornensis*.

8 **Globotruncana helvetica** zone (Total-range zone)
G. imbricata starts together with the guide fossil at the base of the zone, but continues throughout the overlying one. *G. schneegansi* appears slightly above the base of the zone, while *G. sigali* begins only in its upper part; both species continue into overlying zones.
The fact that *Rotalipora* and the Globotruncanidae have never been reported occurring together underlines the importance of the boundary between the *R. cushmani* zone and the *G. helvetica* zone.

9 **Globotruncana schneegansi** zone (Partial-range zone)
That part of the range of the zonal marker, which extends above the top occurrence of *G. helvetica* and below the base occurrence of *G. concavata* constitutes the zone. Furthermore, the top of the zone is marked by the extinction of *G. imbricata*. *G. sigali* may often be abundant throughout this interval, but is not restricted to it. The following species make their first appearance in successively higher levels in the zone: *G. renzi*, *G. angusticarinata*, *G. coronata*, *G. fornicata* and *G. primitiva*.

10 **Globotruncana concavata** zone (Total-range zone)
Forms occurring throughout the zone but not restricted to it are *G. angusticarinata*, *G. coronata* and *G. fornicata*. Forms which become extinct in the basal part are *G. schneegansi*, *G. primitiva* and possibly *G. renzi*. *G. sigali* ranges higher up within this zone. *G. lapparenti* first appears in the lower part, *G. carinata* in the upper part.

11 **Globotruncana carinata** zone (Partial-range zone)
That part of the range of the zonal marker, which extends above the top occurrence of *G. concavata* and below the base occurrence of the single-keeled *G. elevata* and *G. stuartiformis* constitutes the zone. The zonal assemblage includes the longer-ranging *G. angusticarinata*, *G. coronata*, *G. fornicata*, *G. lapparenti* and *G. bulloides*.

12 **Globotruncana elevata** zone (Partial-range zone)
The lower boundary of the zone is characterized by the first appearance of the single-keeled *G. elevata* and *G. stuartiformis*. The upper boundary of the zone coincides with the first occurrences of *G. calcarata* and *G. arca*. Its basal part is well marked by the co-occurrence of *G. elevata* and *G. carinata*.
G. ventricosa, *G. conica* and probably *Praeglobotruncana citae* first appear in the lower part, while *G. angusticarinata* and probably *G. coronata* become extinct in the upper part of the zone.
G. fornicata, *G. lapparenti* and *G. bulloides* are also present throughout the zone.

13 **Globotruncana calcarata** zone (Total-range zone)
Occurring with the characteristic guide fossil but not restricted to this zone are

G. fornicata, *G. lapparenti*, *G. bulloides*, *G. stuartiformis*, *G. conica* and *Praeglobotruncana citae*.

G. arca enters at the base of the zone but continues into the overlying ones. Such species as *G. falsostuarti* and *G. stuarti* make their first appearance immediately after the extinction of *G. calcarata*.

14 **Globotruncana stuartiformis** zone (Concurrent-range zone)

This zone is defined as the range overlap of *G. stuartiformis*, *G. bulloides*, *G. stuarti* and *G. falsostuarti*. The absence in this zone of *G. calcarata*, *G. gansseri*, *G. contusa*, *G. gagnebini* and *Rugoglobigerina rotundata* is an important negative feature.

15 **Globotruncana gansseri** zone (Partial-range zone)

This zone is defined as that part of the range of *G. gansseri* prior to the first appearance of *G. mayaroensis* and *Rugoglobigerina scotti*. *G. contusa*, *G. gagnebini* and *Rugoglobigerina rotundata* make their first appearance at the base of the zone, but range throughout the overlying *G. mayaroensis* zone. Other characteristic forms occurring in the zone, but not restricted to it, include *G. conica*, *G. arca*, *G. stuarti*, *Praeglobotruncana citae*.

16 **Globotruncana mayaroensis** zone (Total-range zone)

In addition to the zonal marker, *Rugoglobigerina scotti* is also restricted to this interval. Other characteristic forms, although not limited to the zone, are *G. conica*, *G. stuarti*, *G. contusa*, *Praeglobotruncana citae* and *Rugoglobigerina rotundata*.

G. gansseri and probably *G. arca* become extinct within this zone.

The entire family Globotruncanidae disappears at the top of the zone.

DESCRIPTIONS AND ILLUSTRATIONS OF
MESOZOIC SPECIES
(arranged in alphabetical order)

Globigerinelloides breggiensis (GANDOLFI)

Reference — *Anomalina breggiensis* GANDOLFI, 1942: Ricerche micropaleontologiche e stratigrafiche sulla Scaglia e sul Flysch cretacici dei dintorni di Balerna (Canton Ticino). — Rivista Italiana di Paleontologia, 48, Suppl. Mem. 4:102, pl. III, fig. 6.

Type locality — Gorge of the Breggia River, northeast of Balerna, near Chiasso, Canton of Ticino, southeastern Switzerland.

Diagnosis — Test planispiral, biumbilicate, slightly evolute, equatorial periphery lobulate.
Wall perforate, surface of the first chambers of the last whorl more or less pitted.
Chambers oblate, the usually 9 chambers of the last whorl increase regularly and slowly in size.
Sutures distinct, depressed, radial.
Primary aperture interiomarginal, an equatorial broad arch with lateral extensions reaching back at either side into the umbilical area bordered by a lip, of which the umbilical parts may be of considerable size (flap-like), the lateral umbilical portions of successive apertures remaining visible as secondary relict apertures.

Strat. distr. — Ranging throughout the *Globigerinelloides breggiensis* zone.

Remarks — Locality of figured specimen is Dyr el Kef section, sample 1F 401, W. Tunisia.

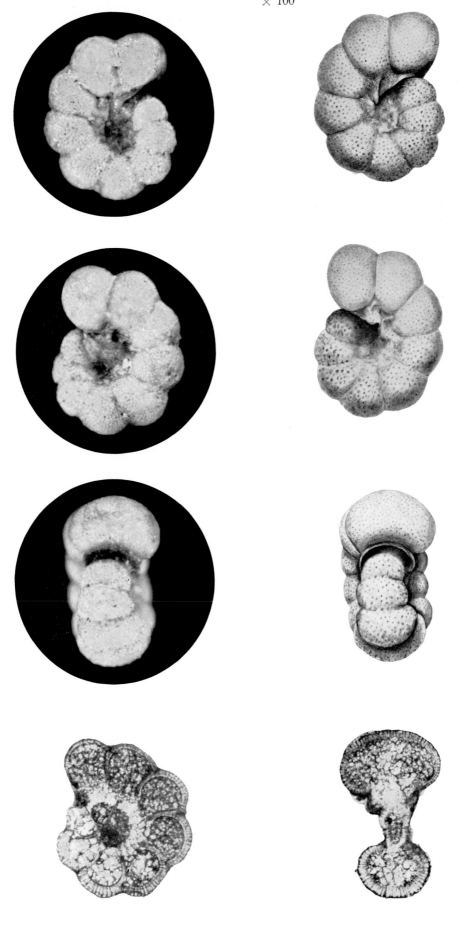

Globotruncana angusticarinata GANDOLFI

Reference   *Globotruncana linnei* (D'ORBIGNY) var. *angusticarinata* GANDOLFI, 1942: Ricerche micropaleontologiche e stratigrafiche sulla Scaglia e sul Flysch cretacici dei dintorni di Balerna (Canton Ticino). — Rivista Italiana di Paleontologia, 48, Suppl. Mem. 4:127, fig. 46, no. 3.

Type locality   Cava della Scabriana, west of Balerna and near the railway line to Lugano, in the vicinity of Chiasso, Canton of Ticino, southeastern Switzerland.

Diagnosis   Test low trochospiral, biconvex; equatorial periphery slightly lobulate, with two beaded keels which become very closely spaced and almost smooth in the last chambers.
Wall perforate, surface smooth.
Chambers angular, compressed, somewhat arched and showing a faint imbricate structure on the spiral side; arranged in about 3 whorls, the 6-7 chambers of the last whorl hardly increase in size.
Sutures on spiral side slightly curved, oblique, raised, beaded, on umbilical side curved, marked by beaded sigmoid septal carinae, which partly border the umbilicus.
Umbilicus rather shallow, wide.
Primary apertures interiomarginal, umbilical, covered by a tegillum.

Strat. distr.   Lower part *Globotruncana schneegansi* zone into upper part *Globotruncana elevata* zone.

Remarks   Locality of figured specimen is Dyr el Kef section, sample 1F 85, W. Tunisia.

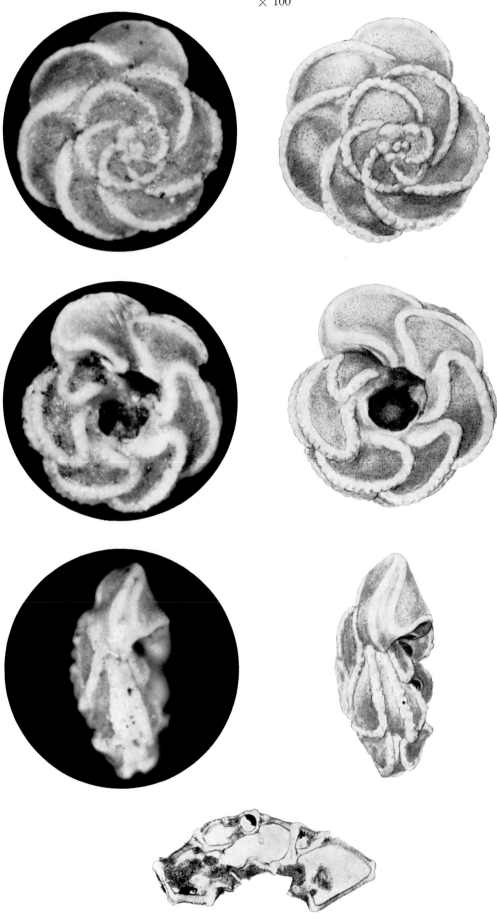

Globotruncana arca (CUSHMAN)

Reference *Pulvinulina arca* CUSHMAN, 1926: Some Foraminifera from the Mendez Shale of Eastern Mexico. — Contr. Cushman Lab. Foram. Res., 2:23, pl. 3, fig. 1a-c.

Type locality Near Huiches, Hacienda El Limon, San Luis Potosí, Mexico.

Diagnosis Test low trochospiral, biconvex; equatorial periphery lobulate with a widely spaced double keel, distinctly beaded except in the last portion.
Wall perforate, surface smooth.
Chambers angular with truncate margins, inflated, arranged in $2\frac{1}{2}$-3 whorls; the 6-7 chambers of the last whorl increase moderately in size.
Sutures on spiral side curved, strongly raised, beaded except for the last chambers; on umbilical side slightly curved, depressed to raised.
Umbilicus fairly deep, wide.
Primary apertures interiomarginal, umbilical, covered by a high tegillum.

Strat. distr. Base *Globotruncana calcarata* zone to top *Globotruncana gansseri* zone. Questionable occurrence in the lower part of the *Globotruncana mayaroensis* zone.

Remarks Locality of figured specimen is Dyr el Kef section, sample 2F 223, W. Tunisia.

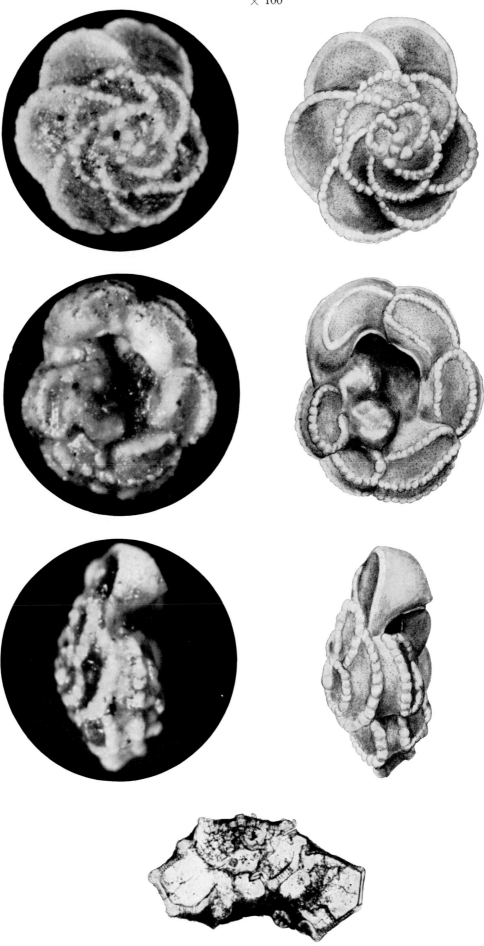

Globotruncana bulloides VOGLER

Reference   *Globotruncana linnei bulloides* VOGLER, 1941: Ober-Jura und Kreide von Misol (Niederländisch-Ostindien). — Palaeontographica, Suppl. IV, Abt. IV:287, pl. XXIII,
figs. 32-39.

Type locality   Island of Misol, east Indonesia.

Diagnosis   Test very low trochospiral, spiral side almost flat to slightly convex, umbilical side
becoming moderately convex; equatorial periphery slightly lobulate to lobulate, with
two widely spaced keels, beaded.
Wall perforate, surface a little rugose in the central part of the chambers, last chambers
usually smooth.
Chambers subangular, strongly inflated, especially the last ones, arranged in about $3\frac{1}{2}$
whorls; the 6-8 chambers of the last whorl increase regularly in size, showing a distinct
imbricate structure on the spiral side.
Sutures on spiral side curved, raised and beaded, the ones between the last three chambers
may be depressed, on umbilical side slightly curved, marked by light, weakly sigmoid,
septal carinae, partly bordering the umbilicus, which may be vague in the last portion of
the test.
Umbilicus shallow, wide.
Primary apertures interiomarginal, umbilical, covered by a tegillum.

Strat. distr.   Upper part *Globotruncana carinata* zone to top *Globotruncana stuartiformis* zone.
Questionable occurrence in lower part *Globotruncana carinata* zone.

Remarks   Locality of figured specimen is Well N. Nederland 12, sample 315-316m.

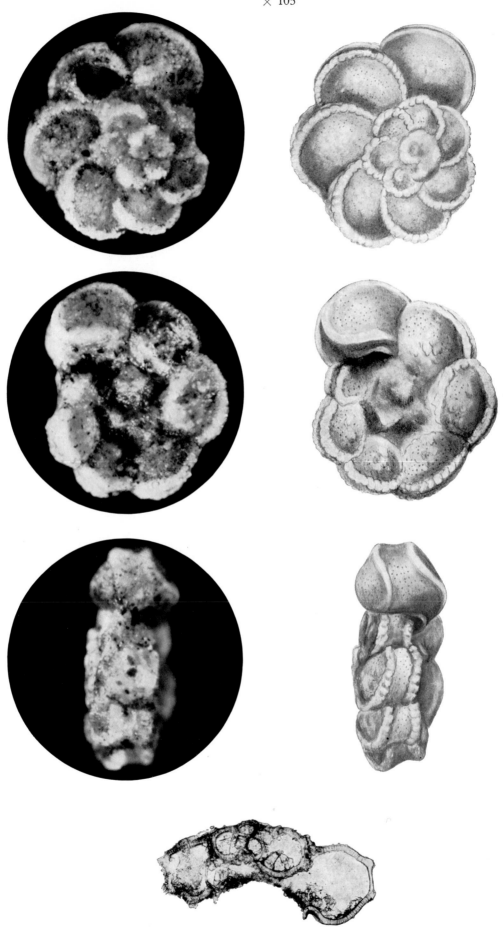

Globotruncana calcarata CUSHMAN

Reference *Globotruncana calcarata* CUSHMAN, 1927: New and interesting Foraminifera from Mexico and Texas. — Contr. Cushman Lab. Foram. Res., 3:115, pl. 23, fig. 10a-b.

Type locality Cut in G.C. and S.F.R.R. at N.edge of Farmerville, Texas, U.S.A.

Diagnosis Test very low trochospiral, spiral side almost flat, umbilical side strongly convex; equatorial periphery stellate, except for the last portion which is rounded, with a distinct single keel which is provided with short spines, one per chamber; keel and spines are beaded, at any rate for the greater part.
Wall perforate, surface rugose, except for the last chambers; degree of rugosity decreases gradually.
Chambers subangular, inflated, arranged in about 3 whorls; the 5-7 chambers of the last whorl increase rather irregularly in size.
Sutures on spiral side slightly curved to almost straight, raised, beaded, on umbilical side radial to slightly curved, depressed to slightly raised, occasionally beaded.
Umbilicus deep, rather narrow to fairly wide.
Primary apertures interiomarginal, umbilical, covered by a tegillum.

Strat. distr. Ranging throughout the *Globotruncana calcarata* zone.

Remarks Locality of figured specimen is sample G 511, W. Tunisia.

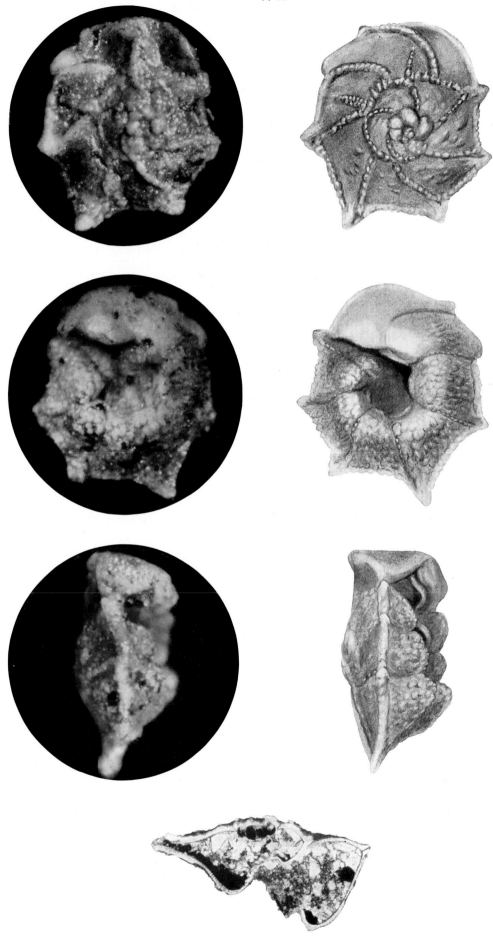

Globotruncana carinata DALBIEZ

Reference   *Globotruncana (Globotruncana) ventricosa carinata* DALBIEZ, 1955: The genus *Globotruncana* in Tunisia. — Micropaleontology, 1 (2):168, text fig. 8.

Type locality   Le Kef-Mellégue, northwestern Tunisia.

Diagnosis   Test very low trochospiral, spiral side most often slightly concave, umbilical side strongly convex; equatorial periphery distinctly lobulate with a rather closely spaced double keel, distinctly beaded, which feature may be absent in the last chamber.
Wall perforate, surface of the first chambers of the last whorl somewhat rugose, last ones smooth.
Chambers angular subconical, moderately inflated, each developing a distinct, usually beaded, carina on top; arranged in 2½-3 whorls, the 5-6 chambers of the last whorl increasing regularly in size.
Sutures on spiral side distinctly curved, in the last whorl raised and beaded, on umbilical side radial, depressed.
Umbilicus deep, wide.
Primary apertures interiomarginal, umbilical, covered by a low tegillum.

Strat. distr.   Upper part *Globotruncana concavata* zone into lowermost part *Globotruncana elevata* zone.

Remarks   This species is characterized by a keel (carina) developed on top of the umbilical side of each chamber, which feature is not present in *Globotruncana concavata* (BROTZEN). Moreover, the chamber has a tendency to become subconical in shape.
It differs from the much younger *Globotruncana gagnebini* TILEV especially in the concavity of the spiral side.
Less striking are the differences from *Globotruncana lamellosa* SIGAL, which develops the same kind of carina, though not so prominent; the chambers, however, are lower and less steep, and the spiral side is slightly convex.
See also remarks on *Globotruncana concavata* (BROTZEN).
Locality of figured specimen is Dyr el Kef section, sample 2F 139, W. Tunisia.

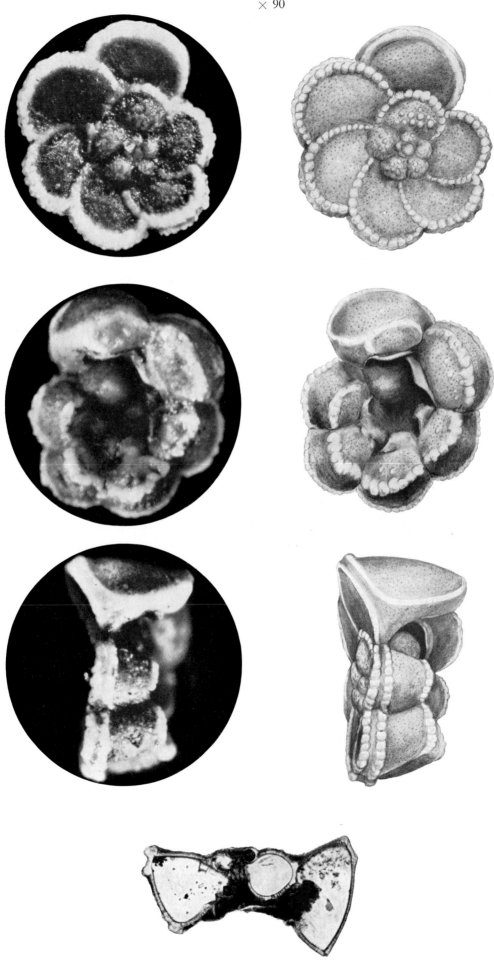

Globotruncana concavata (Brotzen)

Reference *Rotalia concavata* Brotzen, 1934: Foraminiferen aus dem Senon Palästinas. — Zeitschrift des Deutschen Palästina-Vereins, 57:66, pl. 3, fig. b.
Globotruncana (Globotruncana) ventricosa White. Dalbiez, 1955: The genus *Globotruncana* in Tunisia. — Micropaleontology, 1 (2):168, text fig. 7.

Type locality Wadi Madi, on the S. E. flank of Mt. Carmel, Israel.

Diagnosis Test very low trochospiral, spiral side sometimes slightly concave, umbilical side strongly convex; equatorial periphery distinctly lobulate with two closely spaced keels, distinctly beaded except in the last portion.
Wall perforate, surface of the first chambers of the last whorl somewhat rugose, last chambers smooth.
Chambers almost hemispherical, arranged in 2½-3 whorls; the 5-6 chambers of the last whorl increase regularly and usually rapidly in size.
Sutures on spiral side distinctly curved, in last whorl raised and, except in the last chambers, beaded; on umbilical side radial, depressed.
Umbilicus deep, fairly wide.
Primary apertures interiomarginal, umbilical, covered by a low tegillum.

Strat. distr. Ranging throughout *Globotruncana concavata* zone.

Remarks In 1955 Dalbiez described an "evolutionary series" comprising the following three types:
Globotruncana ventricosa primitiva Dalbiez
Globotruncana ventricosa ventricosa White
Globotruncana ventricosa carinata Dalbiez.
After Bolli's study (1957) of the type material of *Globotruncana ventricosa* White, it appeared that the species and subspecies name *ventricosa* of this group had to be changed to *concavata,* a fact which Dalbiez himself had already taken into consideration. For practical purposes (see remarks *Globotruncana lapparenti* Bolli), the following nomenclature is proposed here for the three species concerned:
Globotruncana primitiva Dalbiez
Globotruncana concavata (Brotzen)
Globotruncana carinata Dalbiez.
The main differences between *Globotruncana concavata* and the younger *Globotruncana ventricosa* are:
a. the chambers of the former are hemispherical instead of angular.
b. the height of the chambers increases more rapidly.
c. the surface of the first chambers of the last whorl is somewhat rugose instead of smooth.
d. the tegillum is low instead of high.
Globotruncana asymmetrica Sigal, 1952, is here considered to be a synonym of *Globotruncana concavata.*
Locality of figured specimen is Dyr el Kef section, sample 2F 93, W. Tunisia.

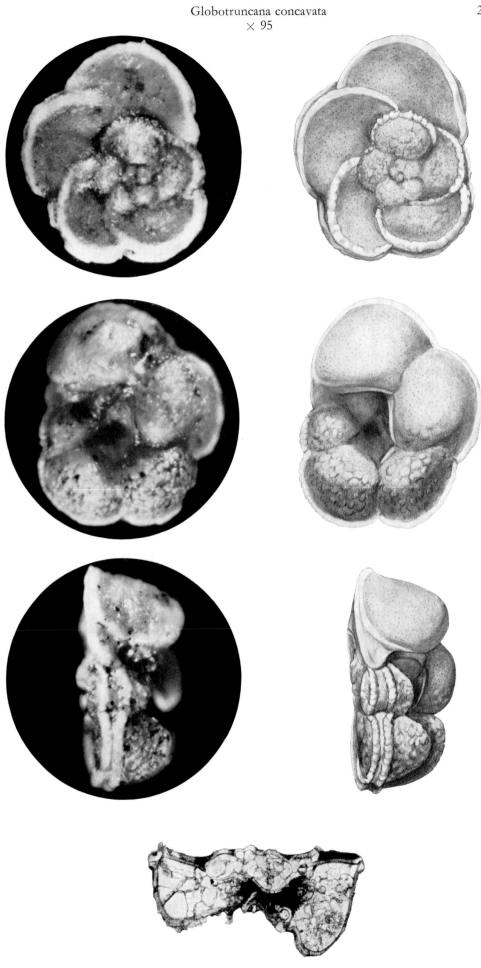

Globotruncana conica WHITE

Reference — *Globotruncana conica* WHITE, 1928: Some index Foraminifera of the Tampico embayment area of Mexico. — Journal of Paleontology, 2 (4):285, pl. 38, fig. 7.

Type locality — Two and two-tenths kilometres east of Guerrero, Tampico embayment, Mexico.

Diagnosis — Test high trochospiral, spiral side strongly convex, umbilical side slightly convex (almost spiroconvex); equatorial periphery slightly lobulate to almost circular, with one keel, moderately beaded except in the last chambers.
Wall perforate, surface smooth.
Chambers subangular to angular, moderately inflated; arranged in $3\frac{1}{2}$-4 whorls, the 6-8 chambers of the last whorl increasing slowly in size.
Sutures on spiral side curved, raised, beaded except in the last chambers, on umbilical side slightly curved, in first part of the last whorl raised and beaded, in later part flush to slightly depressed.
Umbilicus deep, wide.
Primary apertures interiomarginal, umbilical, covered by a high tegillum.

Strat. distr. — Lower part *Globotruncana elevata* zone to top *Globotruncana mayaroensis* zone.

Remarks — In the same paper WHITE describes a variety *Globotruncana conica* var. *plicata* as "being fluted or having folds on the dorsal side, causing the outline as seen from above to appear fluted or polygonal". This variety is a synonym of *Globotruncana contusa* CUSHMAN.
Locality of figured specimen is Dyr el Kef section, sample 2F 227, W. Tunisia.

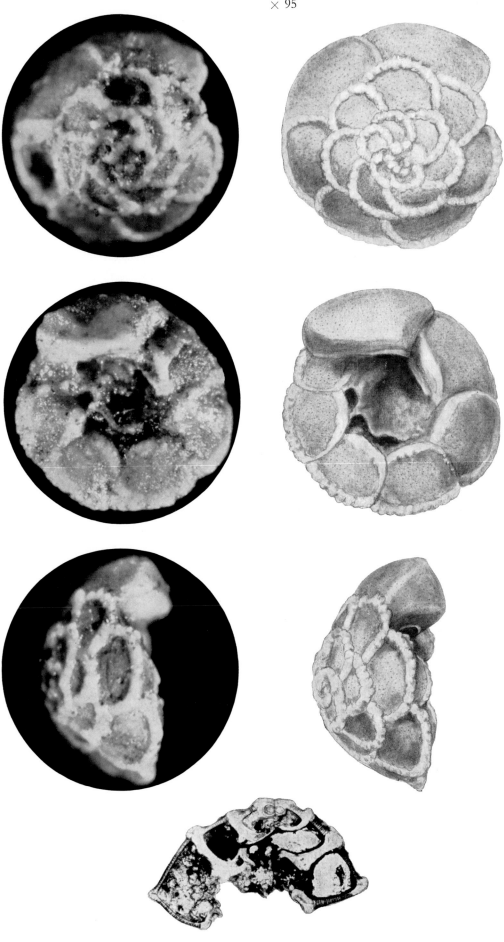

Globotruncana contusa (CUSHMAN)

Reference   *Pulvinulina arca* var. *contusa* CUSHMAN, 1926: Some Foraminifera from the Mendez shale of eastern Mexico. — Contr. Cushman Lab. Foram. Res., 2:23, no figure.

Type locality   Near Coco, Hacienda El Limon, San Luis Potosí, Mexico.

Diagnosis   Test high trochospiral; spiral side strongly convex, umbilical side flattened to slightly concave (spiroconvex); equatorial periphery polygonal with 2 keels, distinctly beaded, except sometimes in the last chamber.
Wall perforate, surface smooth.
Chambers angular with truncate margins, distinctly concave on the spiral side, which gives the test the appearance of a more or less pyramidal form; arranged in 3½-4 whorls, the 5-7 chambers of the last whorl increasing slowly and rather irregularly in size.
Sutures on spiral side slightly curved to arched, strongly oblique, raised, distinctly beaded, on umbilical side slightly curved to radial, raised and beaded to sometimes depressed in the last chambers of the final whorl.
Umbilicus deep, wide.
Primary apertures interiomarginal, umbilical, covered by a relatively low tegillum.

Ranging throughout the *Globotruncana gansseri* zone and the *Globotruncana mayaroensis* zone.

Strat. distr.   Text figure 2 in DE LAPPARENT's publication of 1918 shows a number of sections of different types of *Rosalina linnei* D'ORBIGNY. One of these is defined as a "section de la mutation caliciforme". In fact it is a section of what is now generally thought to be *Globotruncana contusa*.

Remarks   Locality of figured specimen is Dyr el Kef section, sample 2F 224, W. Tunisia.

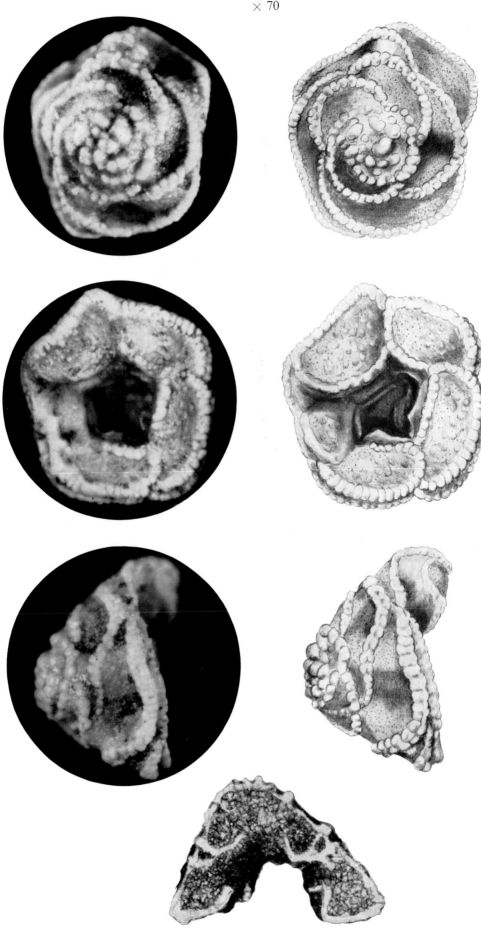

Globotruncana coronata BOLLI

Reference   *Globotruncana lapparenti coronata* BOLLI, 1944: Zur Stratigraphie der Oberen Kreide in den höheren helvetischen Decken. — Eclogae Geologicae Helvetiae, 37 (2):233, fig. 1, no. 21, 22; pl. IX, figs. 14, 15.

Type locality   Between Wildhaus and Voralpsee, Säntis area, northeastern Switzerland.

Diagnosis   Test low trochospiral, biconvex; equatorial periphery lobulate with two closely spaced keels, distinctly beaded, at least in the first part.
Wall perforate, surface smooth.
Chambers angular, strongly compressed, arranged in 2½-3 whorls; the 5-7 chambers of the last whorl increase regularly in size; early whorls small by comparison.
Sutures on spiral side curved, raised, beaded, on umbilical side curved, marked by beaded sigmoid septal carinae, which partly border the umbilicus.
Umbilicus shallow, wide.
Primary apertures interiomarginal, umbilical, covered by a tegillum.

Strat. distr.   Upper part *Globotruncana schneegansi* zone to upper part *Globotruncana carinata* zone. Questionable occurrences in lower part *Globotruncana schneegansi* zone, uppermost *Globotruncana carinata* zone and *Globotruncana elevata* zone.

Remarks   See remarks *Globotruncana lapparenti* BOLLI.
Locality of figured specimen is Dyr el Kef section, sample 2F 139, W. Tunisia.

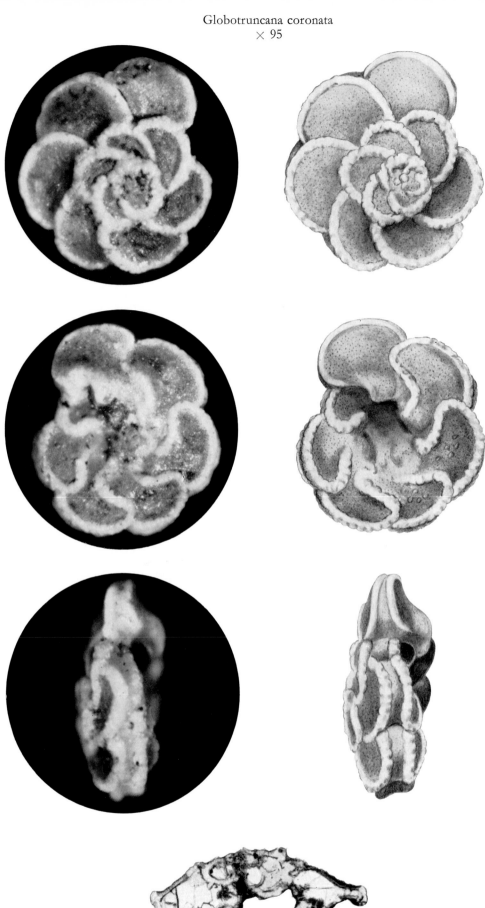

Globotruncana elevata (BROTZEN)

Reference   *Rotalia elevata* BROTZEN, 1934: Foraminiferen aus dem Senon Palästinas. — Zeitschrift des Deutschen Palästina-Vereins, 57:66, pl. 3, fig. c.
Globotruncana (Globotruncana) elevata elevata (BROTZEN): The genus *Globotruncana* in Tunisia. — Micropaleontology, 1 (2):169, text fig. 9 a-c.

Type locality   Wadi Madi, on the S. E. flank of Mt. Carmel, Israel.

Diagnosis   Test very low trochospiral, central part of spiral side slightly convex to convex, spiral side of last whorl flat to slightly concave, umbilical side strongly convex; equatorial periphery lobulate to slightly lobulate, with one keel, moderately beaded except in the last chambers.
Wall perforate, surface smooth.
Chambers subangular to angular, moderately inflated, sometimes slightly overlapping, with a kind of carina on top of each chamber as continuation of the partly raised sutures of the umbilical side; arranged in about 3 whorls, the usually 6-8 chambers of the last whorl increasing regularly in size.
Sutures on spiral side distinctly curved, at least in the last chambers, raised, moderately beaded; on umbilical side slightly curved, in first part of the last whorl raised and beaded, in later part slightly depressed.
Umbilicus deep, wide.
Primary apertures interiomarginal, umbilical, covered by a tegillum.

Strat. distr.   Base *Globotruncana elevata* zone into lower part *Globotruncana calcarata* zone. Questionable occurrence in upper part *Globotruncana calcarata* zone.

Remarks   The *Globotruncana elevata* group is another evolutionary series established by DALBIEZ, containing some single-keeled *Globotruncana* types of the Campanian/Maastrichtian.
DALBIEZ recommended a complete revision of the "group *elevata-rosetta-stuarti*", but in the meantime proposed the following nomenclatural modifications:
stage 1 - *Globotruncana elevata elevata* (BROTZEN),
stage 2 - *Globotruncana elevata stuartiformis* DALBIEZ,
stage 3 - *Globotruncana stuarti* (DE LAPPARENT).
For practical reasons (see remarks *Globotruncana lapparenti* BOLLI) the following nomenclature is proposed here:
        Globotruncana elevata (BROTZEN),
        Globotruncana stuartiformis DALBIEZ,
        Globotruncana stuarti (DE LAPPARENT).
Globotruncana andori DE KLASZ, 1953, appears to be closely related to this group.
Globotruncana rosetta (CARSEY), 1926, may be another species related to the above mentioned group. Its stratigraphic distribution is within the potential range of the group.
Locality of figured specimen is Dyr el Kef section, sample 2F 171, W. Tunisia.

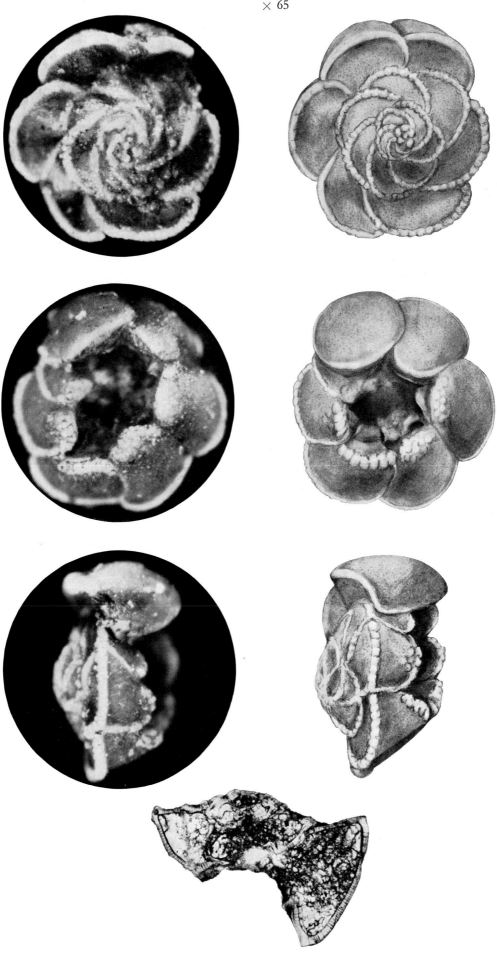

Globotruncana falsostuarti SIGAL

Reference   *Globotruncana falsostuarti* SIGAL, 1952: Aperçu stratigraphique sur la micropaléontologie du Crétacé. — 19e Congrès Géologique International, Monographies Régionales, Ser. 1 (26) :43, text fig. 46.

Type locality   Not given (probably northern Algeria).

Diagnosis   Test low trochospiral, biconvex; equatorial periphery lobulate with two keels, which are fused in the central part of each chamber; beaded, which feature may be absent in the last chambers.
Wall perforate, surface smooth.
Chambers subangular to angular with partly truncate margins, moderately inflated, slightly overlapping, arranged in 3-3½ whorls, the 7-8 chambers of the last whorl increasing regularly and slowly in size.
Sutures on spiral side curved to slightly curved, raised, beaded, on umbilical side curved, raised, beaded, at least for the greater part.
Umbilicus fairly deep, wide.
Primary apertures interiomarginal, umbilical, covered by a high tegillum.

Strat. distr.   Ranging throughout *Globotruncana stuartiformis* zone, *Globotruncana gansseri* zone and *Globotruncana mayaroensis* zone.

Remarks   The original description of this species is insufficient, but through oral information from the author it appears that the keel pattern, as described above, is the most characteristic feature of the species, and is clearly visible in the author's figure of the holotype. The degree of fusing is variable; the specimen shown here exhibits a lesser degree of fusing of keels than is seen in the holotype.
It is possible that *Globotruncana falsostuarti* may be related to, or even synonymous with, *Globotruncana leupoldi* BOLLI, 1944.
Locality of figured specimen is Dyr el Kef section, sample 2F 218, W. Tunisia.

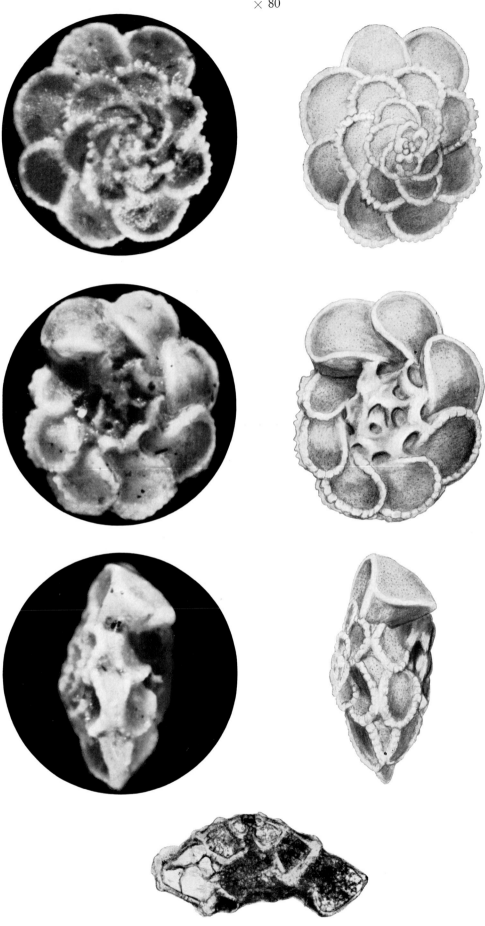

Globotruncana fornicata PLUMMER

Reference *Globotruncana fornicata* PLUMMER, 1931: Some Cretaceous Foraminifera in Texas. — The University of Texas Bulletin, 3101:198, pl. XIII, figs. 4-6.

Type locality Station 226-T-8, on right bank of Onion Creek near bridge at Moore and Berry's Crossing, eight and onehalf miles in a straight line southeast of the capital Austin, Travis County, Texas, U.S.A.

Diagnosis Test fairly high trochospiral, biconvex; equatorial periphery slightly lobulate, with two keels, moderately beaded.
Wall perforate, surface smooth.
Chambers angular with truncate margins, long, strongly arched, inflated inner parts on the spiral side, arranged in 2½-3 whorls; the 4-5 chambers of the last whorl increase rapidly in size.
Sutures on spiral side strongly oblique and slightly curved, raised, except for the inner whorl, beaded; on umbilical side strongly curved, depressed to flush, partly marked by beaded continuations of one of the keels.
Umbilicus fairly deep, wide.
Primary apertures interiomarginal, umbilical, covered by a tegillum.

Strat. distr. Upper part *Globotruncana schneegansi* zone to top *Globotruncana stuartiformis* zone. Questionable occurrence in lower part *Globotruncana gansseri* zone.

Remarks As PLUMMER stated, this species is "easily distinguished from its congeners by its narrow and strongly arched chambers that sweep in strong curves on the dorsal face".
Locality of figured specimen is Dyr el Kef section, sample 2F 171, W. Tunisia.

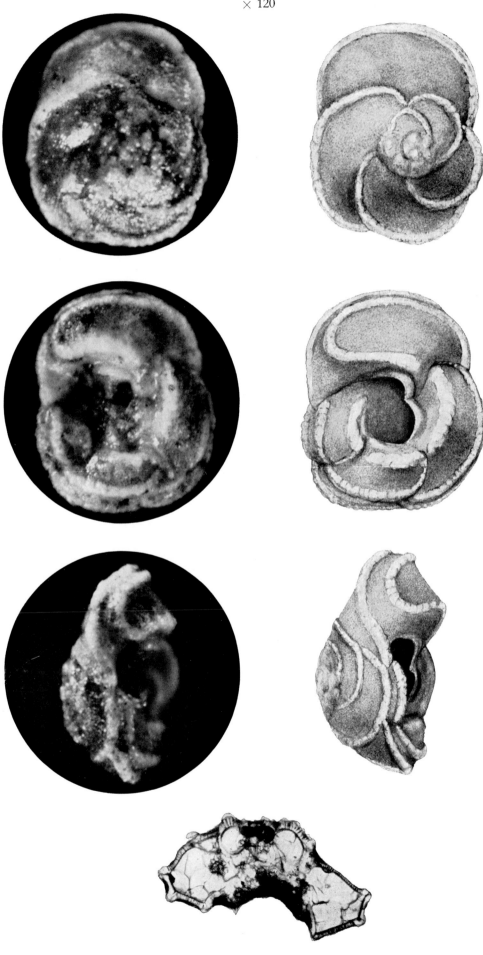

Globotruncana gagnebini TILEV

Reference   *Globotruncana gagnebini* TILEV, 1951: Etude des Rosalines maastrichtiennes (genre *Globotruncana*) du Sud-Est de la Turquie (sondage de Ramandağ). — Publications de l'Institut d'Etudes et de Recherches Minières de Turquie, B (16):50, text fig. 14.

Type locality   Well Ramandağ IV, 1,170m, S. E. Turkey.

Diagnosis   Test very low trochospiral, spiral side flat or slightly convex, umbilical side strongly convex; equatorial periphery distinctly lobulate with two rather closely speced keels which may approach each other or even fuse in the last chamber; beaded, except for the last chamber.
Wall perforate, surface rugose, degree of rugosity decreases towards the last chamber, which is generally smooth; tendency to develop a carina on top of the chambers increases towards the last chamber.
Chambers subangular, inflated on both umbilical side and spiral side, arranged in $2\frac{1}{2}$-3 whorls, the 4-5 chambers of the last whorl increasing rapidly in size.
Sutures on spiral side curved, raised, beaded; on umbilical side radial, depressed.
Umbilicus deep, fairly wide.
Primary apertures interiomarginal, umbilical, covered by a tegillum.

Strat. distr.   Ranging throughout *Globotruncana gansseri* zone and *Globotruncana mayaroensis* zone.

Remarks   TILEV describes *Globotruncana gagnebini* as a very variable form. Other illustrated specimens of his species from higher samples of the same well appear to be very similar to or even identical with *Globotruncana lamellosa* SIGAL and *Globotruncana aegyptiaca* NAKKADY.
Locality of figured specimen is well Fahud-1, sample 145', Oman.

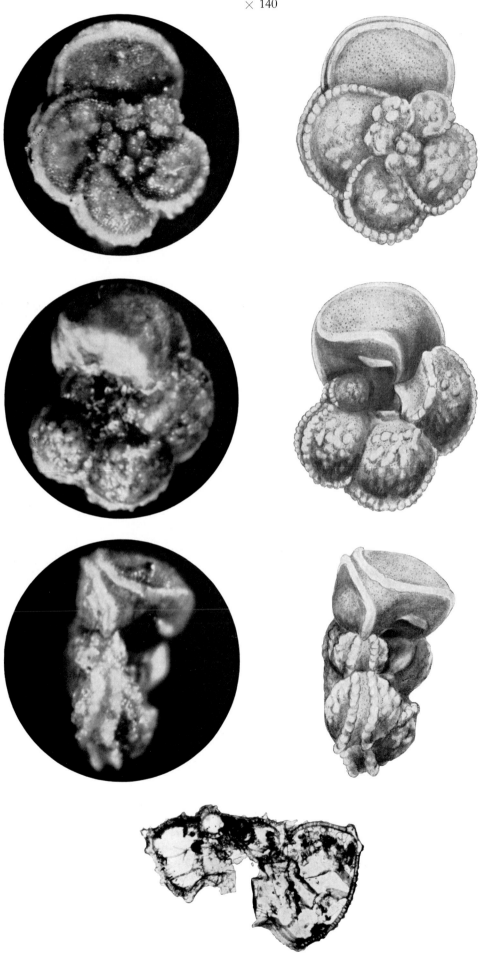

Globotruncana gansseri BOLLI

Reference   *Globotruncana gansseri* BOLLI, 1951 : The genus *Globotruncana* in Trinidad, B.W.I. — Journal of Paleontology, 25 (2) :196, pl. 35, figs. 1-3.

Type locality   Subsurface section in the Brighton area near Pitch Lake, southwestern Trinidad.

Diagnosis   Test very low trochospiral, spiral side flat, umbilical side strongly convex; equatorial periphery slightly lobulate to almost circular, with one keel, beaded, at least in the first part.
Wall perforate, surface of the umbilical side rugose, degree of rugosity decreases toward the last chambers, which are smooth; surface of the spiral side smooth, except the initial part.
Chambers almost hemispherical, arranged in 2½-3 whorls; the 5-6 chambers of the last whorl increase regularly in size.
Sutures on spiral side curved, raised in the last whorl, lightly beaded; on umbilical side the first ones radial, the last ones slightly curved, depressed.
Umbilicus deep, wide.
Primary apertures interiomarginal, umbilical, covered by a tegillum.

Strat. distr.   Base *Globotruncana gansseri* zone to upper part *Globotruncana mayaroensis* zone.

Remarks   *Globotruncana lugeoni* TILEV, 1951 is a synonym of *Globotruncana gansseri*.
Locality of figured specimen is sample St. 1289, Turkey.

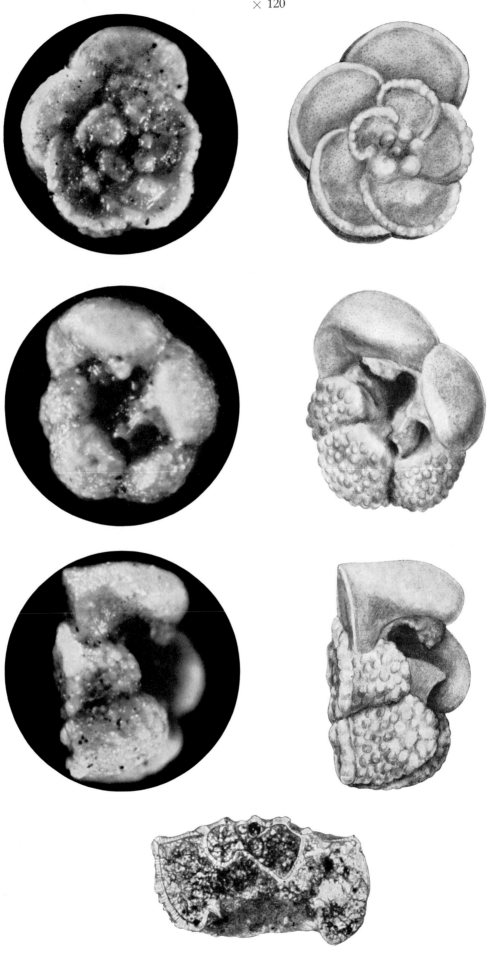

Globotruncana helvetica BOLLI

Reference   *Globotruncana helvetica* BOLLI, 1945: Zur Stratigraphie der Oberen Kreide in den höheren helvetischen Decken. — Eclogae Geologicae Helvetiae, 37 (2):226, pl. IX, fig. 6.

Type locality   Sample 952, 15-20 metres above the top of the "Knollenschichten" in the Säntis section, between the meteorological station and the old hotel on the crest of the Säntis Range, Canton of St. Gall, eastern Switzerland.

Diagnosis   Test very low trochospiral, spiral side almost flat, inner whorls often slightly raised, umbilical side strongly inflated; equatorial periphery lobulate with one keel, which may be weakened in the last chamber, lightly beaded.

Wall perforate; surface on umbilical side distinctly rugose, the last chamber to a lesser extent; on spiral side lightly rugose.

Chambers hemispherical, arranged in about 3 whorls, the 5-6 chambers increasing regularly in size; on spiral side a tendency to develop an imbricate structure.

Sutures on spiral side curved, raised and beaded, on umbilical side almost radial, depressed.

Umbilicus fairly deep, wide.

Primary apertures interiomarginal, umbilical, covered by a simple, low tegillum.

Strat. distr.   Ranging throughout *Globotruncana helvetica* zone.

Remarks   Locality of figured specimen is Dyr el Kef section, sample 2F 40, W. Tunisia.

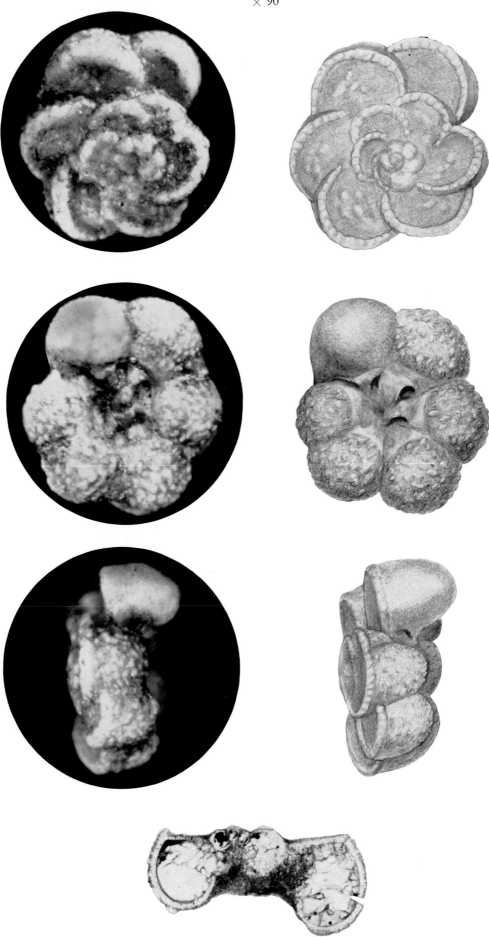

Globotruncana imbricata MORNOD

Reference *Globotruncana imbricata* MORNOD, 1949: Les Globorotalidés du Crétacé supérieur du Montsalvens (Préalpes fribourgeoises). — Eclogae Geologicae Helvetiae, 42 (2):589, text fig. 5, III a-d.

Type locality Middle of Profile III, at about 975 metres above sea level in the Ruisseau des Covayes, on the southeastern slope of the Montsalvens chain, north of Cerniat in the Préalpes fribourgeoises, Switzerland.

Diagnosis Test low trochospiral, spiral side convex, umbilical side becoming concave; equatorial periphery slightly lobulate to lobulate with two beaded keels, which are absent or at least weakened in the ultimate and penultimate chambers.
Wall perforate, surface of the chambers, except the last ones, may be slightly rugose.
Chambers subangular, slightly inflated, arranged in 2½-3 whorls, the 5-6 chambers of the last whorl increasing regularly and slowly in size, showing on the spiral side an imbricate structure (like roof tiles).
Sutures on spiral side curved, raised and beaded, except the one between ultimate and penultimate chamber, which is commonly depressed; on umbilical side radial, depressed. Umbilicus shallow, fairly wide.
Primary apertures interiomarginal, umbilical, covered by a simple low tegillum.

Strat. distr. Ranging throughout *Globotruncana helvetica* zone and *Globotruncana schneegansi* zone.

Remarks Is is not inadvisable, to allow for a fairly large degree of variation in this species. The variants differ from each other only in minor characteristics, while the most important features—the imbricate structure of the chambers and the typical position of the last chamber—are always present.
From this point of view *Globotruncana imbricata* and *Globotruncana inflata* BOLLI, 1944, may be considered as synonymous. Priority, however, should be given to *Globotruncana imbricata* MORNOD, which has been described from isolated specimens.
Locality of figured specimen is Dyr el Kef section, sample 2F 62, W. Tunisia.

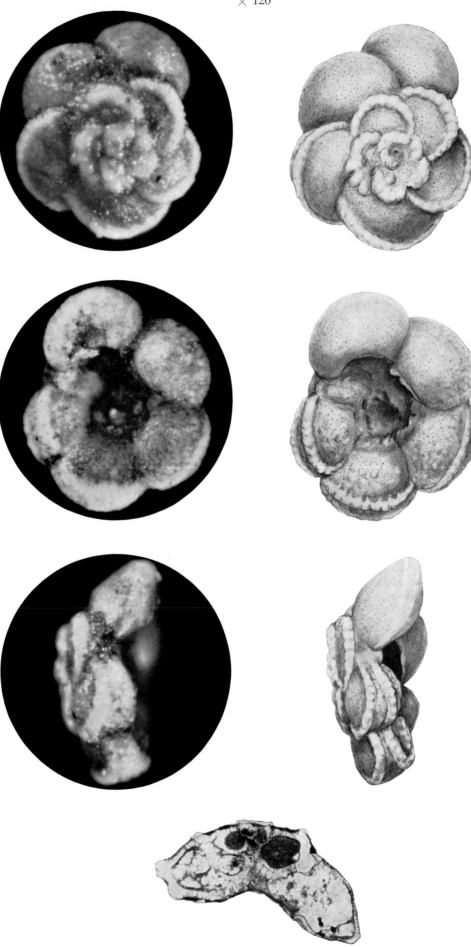

Globotruncana lapparenti BOLLI

Reference *Globotruncana lapparenti lapparenti* BOLLI, 1944: Zur Stratigraphie der Oberen Kreide in den höheren helvetischen Decken. — Eclogae Geologicae Helvetiae, 37 (2):230, text fig. 1, nos. 15, 16; pl. IX, fig. 1.

Type locality Sample no. 1384, about 31.5 metres above the base of the Gatter section, Säntis area, northeastern Switzerland.

Diagnosis Test very low trochospiral, spiral side almost flat to slightly convex, umbilical side practically flat; equatorial periphery slightly lobulate to lobulate with two widely spaced keels, beaded.
Wall perforate, surface smooth.
Chambers almost rectangular, may be slightly inflated, arranged in about 3½ whorls, the 5-8 chambers of the last whorl increasing regularly in size.
Sutures on spiral side curved, oblique, strongly raised, beaded; on umbilical side curved, marked by beaded sigmoid septal carinae, which partly border the umbilicus.
Umbilicus fairly shallow, wide.
Primary apertures interiomarginal, umbilical, covered by a tegillum.

Strat. distr. Lower part *Globotruncana concavata* zone to uppermost part *Globotruncana stuartiformis* zone.

Remarks In 1936, with reference to the species which he had determined and figured as *Globotruncana ventricosa* WHITE, BROTZEN expressed a view which also has a bearing on the *Globotruncana linneiana* (D'ORBIGNY) problem. He said that he could distinguish *Globotruncana linneiana* D'ORBIGNY; *Globotruncana marginata* (REUSS); *Globotruncana canaliculata* (REUSS); *Globotruncana canaliculata* var. *ventricosa* WHITE; and, furthermore, his new species *Globotruncana lapparenti*.
For this last, BROTZEN chose as type species the *Rosalina linnei* D'ORBIGNY group of forms of DE LAPPARENT (1918), consisting of six types, as a result of which the definition of the species is very elastic. (According to BROTZEN; one of the characters in which these six types differ from *Globotruncana linneiana* D'ORBIGNY is in the pattern of the sutures on the umbilical side.)
In 1941 VOGLER made use of a ternary nomenclature, but he, on the other hand, looked upon *Globotruncana linnei* D'ORBIGNY (= *G. linneiana* D'ORBIGNY) as "Grossart"
In 1944 BOLLI also went a step farther and, after removing type 6, "la mutation caliciforme" (= *Globotruncana contusa* (CUSHMAN)), from the group, chose no. 1 of these types of DE LAPPARENT's—viz. the subspecies *Globotruncana lapparenti lapparenti* BOLLI—as "zentrale Typus der Grossart *Globotruncana lapparenti* BROTZEN"
For practical reasons the subspecies *Globotruncana lapparenti lapparenti* BOLLI, 1944, which is described here, is given as the species *Globotruncana lapparenti* BOLLI.
Other amendments of the kind are as follows:
Globotruncana lapparenti bulloides VOGLER, 1941 = *G. bulloides* VOGLER
Globotruncana lapparenti inflata BOLLI, 1944 = *G. inflata* BOLLI (See remarks *G. imbricata*)
Globotruncana lapparenti coronata BOLLI, 1944 = *G. coronata* BOLLI
Globotruncana lapparenti tricarinata (QUEREAU), 1893 = *G. tricarinata* QUEREAU
Locality of figured specimen is Dyr el Kef section, sample 2F 95, W. Tunisia.

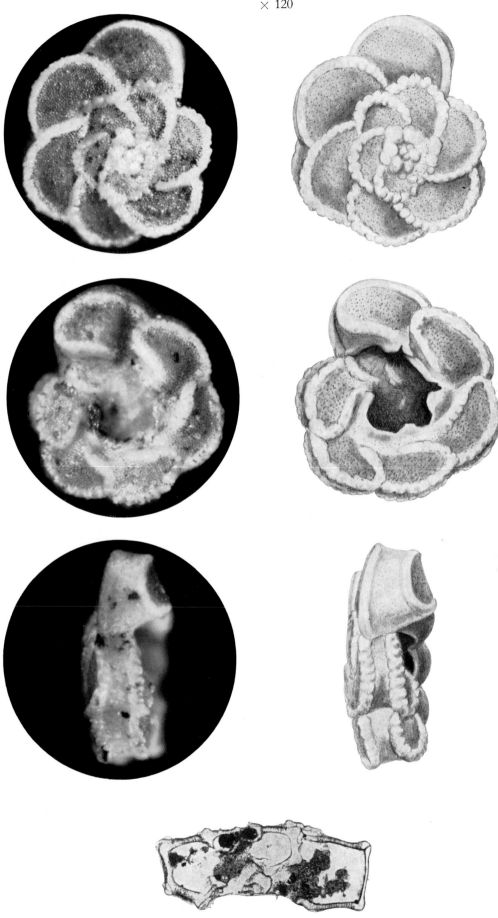

Globotruncana mayaroensis BOLLI

Reference *Globotruncana mayaroensis* BOLLI, 1951: The genus *Globotruncana* in Trinidad, B.W.I. —Journal of Paleontology, 25 (2):198, pl. 35, figs. 10-12.

Type locality Subsurface section in the Guayaguayare area, County of Mayaro, southeastern Trinidad.

Diagnosis Test very low trochospiral, spiral side almost flat to slightly convex, umbilical side moderately concave; equatorial periphery lobulate with two beaded keels, of which the one on the umbilical side becomes strongly arched towards the last chamber.
Wall perforate, surface ornamented with fine nodes, including the sidewall between the keels.
Chambers angular-truncate, on umbilical side more inflated than on spiral side, arranged in about 3 whorls, the 4-6 (usually 5) chambers of the last whorl increasing sometimes rapidly in size; on the spiral side a tendency to develop an imbricate structure.
Sutures on spiral side curved, raised and beaded, on umbilical side radial, depressed.
Umbilicus shallow, fairly wide.
Primary apertures interiomarginal, umbilical, covered by a tegillum.

Strat. distr. Ranging throughout *Globotruncana mayaroensis* zone.

Remarks In 1957 (U.S. Bull. 215) BOLLI, LOEBLICH and TAPPAN introduced the new genus *Abathomphalus,* with *Globotruncana mayaroensis* as its type species. But since the essential generic characteristics are the same-as became obvious after several specimens had been thoroughly cleaned-preference has again been given to the genus name *Globotruncana.*
Locality of figured specimen is sample 3K 235, W. Tunisia.

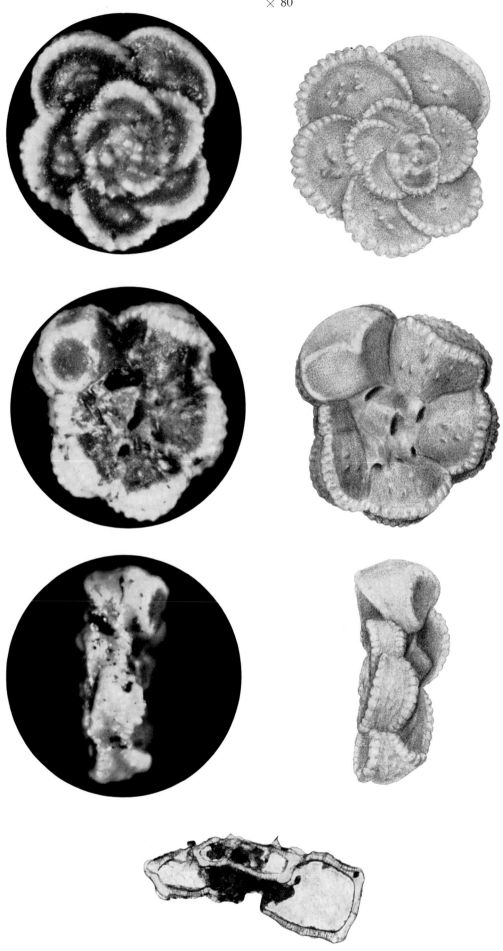

Globotruncana primitiva DALBIEZ

Reference *Globotruncana (Globotruncana) ventricosa primitiva* DALBIEZ, 1955: The genus *Globotruncana* in Tunisia.' — Micropaleontology, 1 (2):168, text fig. 6.

Type locality Le Kef-Mellegue, northwestern Tunisia.

Diagnosis Test very low trochospiral, spiral side almost flat, umbilical side convex; equatorial periphery lobulate with two very closely spaced keels, beaded, which feature may be absent in the last chamber.

Wall perforate, surface of the first chambers of the last whorl somewhat rugose, last chambers smooth.

Chambers subangular, moderately inflated, arranged in about 3 whorls; the (generally 6) chambers of the last whorl increase regularly in size.

Sutures on spiral side distinctly curved, in last whorl raised and beaded; on umbilical side radial, depressed.

Umbilicus shallow, wide.

Primary apertures interiomarginal, umbilical, covered by a low tegillum.

Strat. distr. Upper part *Globotruncana schneegansi* zone and lower part *Globotruncana concavata* zone.

Remarks See remarks *Globotruncana concavata* (BROTZEN).
Locality of figured specimen is sample C 253, W. Tunisia.

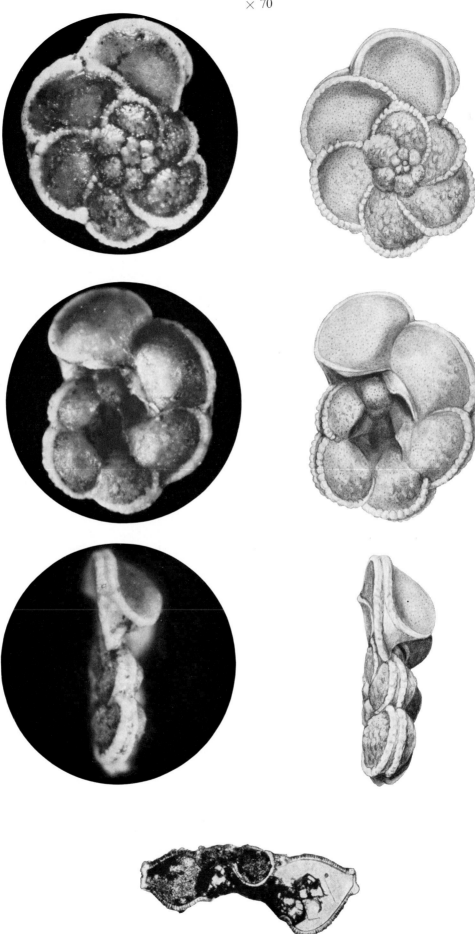

Globotruncana renzi GANDOLFI

Reference *Globotruncana renzi* GANDOLFI, 1942: Ricerche micropaleontologiche e stratigrafiche sulla Scaglia e sul Flysch cretacici dei dintorni di Balerna (Canton Ticino). — Rivista Italiana di Paleontologia, 48, Suppl. Mem. 4:124, text fig. 45, pl. III, fig. 1a-c.

Type locality Gorge of the Breggia River, northeast of Balerna, near Chiasso, Canton of Ticino, south-eastern Switzerland.

Diagnosis Test low trochospiral, biconvex; equatorial periphery slightly lobulate to lobulate, with two closely spaced beaded keels in early portion of last whorl, last portion (ultimate and occasionally penultimate chamber) with one smooth keel only.
Wall perforate, surface smooth, except on the umbilical side of the first chambers of the last whorl, which may be slightly rugose.
Chambers angular, compressed to slightly inflated, the 5-6 chambers of the last whorl increasing somewhat irregularly in size.
Sutures on spiral side curved to oblique, in last whorl raised and beaded; on umbilical side slightly curved, raised and beaded.
Umbilicus shallow, wide.
Primary apertures interiomarginal, umbilical, covered by a simple low tegillum.

Strat. distr. Almost throughout *Globotruncana schneegansi* zone, questionable occurrence in lower part *Globotruncana concavata* zone.

Remarks *Globotruncana renzi* is probably an intermediate form between *Globotruncana schneegansi* SIGAL and representatives of the "*Globotruncana lapparenti* group".
It may be mentioned that the thin section presented by GANDOLFI on plate X of his paper (1942) is not a *Globotruncana renzi*, but is most probably a section of *Globotruncana imbricata* MORNOD.
Locality of figured specimen is Dyr el Kef section, sample 2F 45, W. Tunisia.

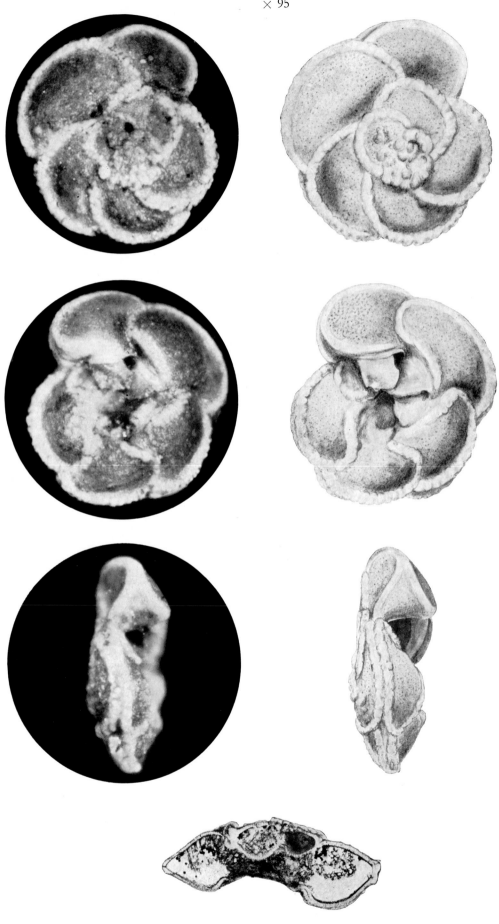

Globotruncana schneegansi SIGAL

Reference *Globotruncana schneegansi* SIGAL, 1952: Aperçu stratigraphique sur la micropaléontologie du Crétacé. — 19e Congrès Géologique International, Monographies Régionales, 1 (26):33, fig. 34.

Type locality Not given (probably northern Algeria).

Diagnosis Test low trochospiral, biconvex; equatorial periphery distinctly lobulate with one beaded keel, which may be weakened and smooth in the last chamber.
Wall perforate, surface slightly rugose.
Chambers angular, the first ones of the last whorl slightly compressed, the last ones slightly inflated; a carina, as continuation of the sutures, may be present on top of all chambers except the last one, arranged in about 3½ whorls; the 5-6 chambers of the last whorl increase somewhat irregularly in size.
Sutures on spiral side curved to oblique, raised, beaded; on umbilical side slightly curved to almost radial, raised and beaded to flush in the first portion of the last whorl, depressed in the last portion.
Umbilicus fairly deep, wide.
Primary apertures interiomarginal, umbilical, covered by a simple, low tegillum.

Strat. distr. Lower part *Globotruncana helvetica* zone into lowermost part *Globotruncana concavata* zone.

Remarks Locality of figured specimen is sample 3K 55, W. Tunisia.

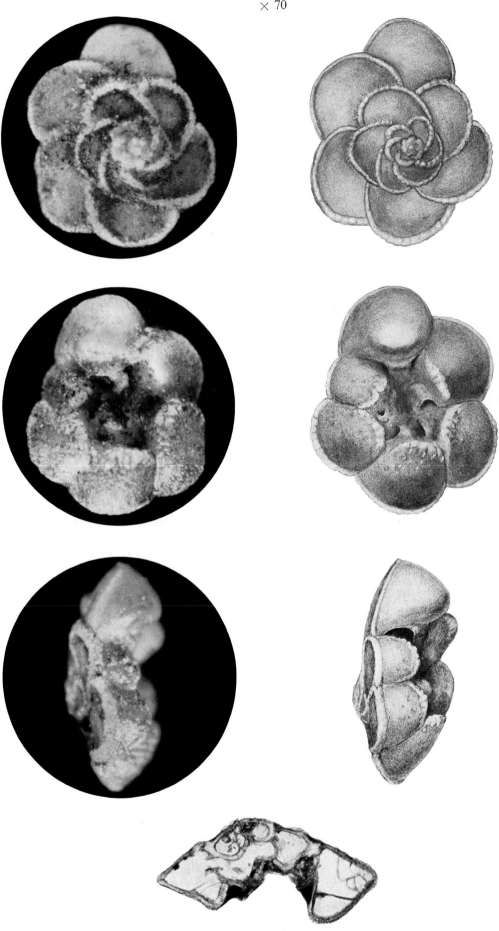

Globotruncana sigali REICHEL

Reference *Globotruncana (Globotruncana) sigali* REICHEL, 1950: Observations sur les *Globotruncana* du gisement de la Breggia (Tessin). — Eclogae Geologicae Helvetiae, 42 (2):610, pl. 16, fig. 7.

Type locality Sidi Aissa, south of Aumale, Algeria.

Diagnosis Test low trochospiral, biconvex; equatorial periphery moderately lobulate with one slightly beaded keel, which is almost smooth in the last chamber.
Wall perforate, surface smooth.
Chambers angular, compressed; arranged in about $3\frac{1}{2}$ whorls, the 5-7 (usually 6) chambers of the last whorl increasing regularly in size.
Sutures on spiral side raised, beaded, slightly curved to almost straight, oblique, in the last whorl becoming more radial, on umbilical side curved, marked by beaded sigmoid, septal carinae, partly bordering the umbilicus.
Umbilicus shallow, moderately wide.
Primary apertures interiomarginal, umbilical, covered by a simple, low tegillum.

Strat. distr. Uppermost part *Globotruncana helvetica* zone into upper part *Globotruncana concavata* zone.

Remarks Locality of figured specimen is Dyr el Kef section, sample 2F 42, W. Tunisia.

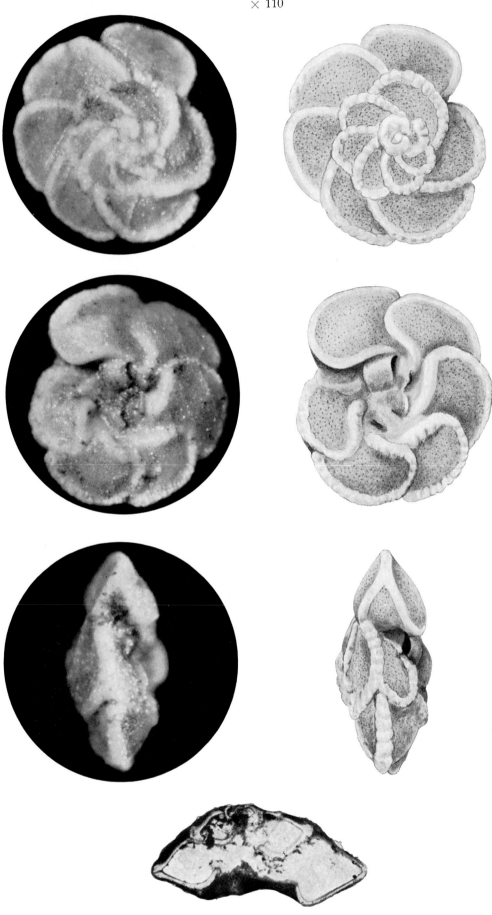

Globotruncana stuarti (De Lapparent)

Reference *Rosalina stuarti* De Lapparent, 1918: Etude lithologique des terrains crétacés de la région d'Hendaye. — Mémoires pour servir á l'explication de la carte géologique détaillée de la France: 11, text fig. 4.

Type locality Pointe Sainte-Anna, Hendaye area, western Pyrenees, France.

Diagnosis Test low trochospiral, biconvex; equatorial periphery slightly lobulate to almost circular, with one beaded keel.
Wall perforate, surface smooth.
Chambers angular, slightly overlapping, with a beaded carina along the umbilical area as continuation of the raised sutures; arranged in about 3½ whorls, the usually 6-7 chambers of the last whorl hardly increasing in size.
Sutures on spiral side straight to slightly curved, raised, moderately beaded; on umbilical side curved, raised, lightly beaded.
Umbilicus deep, wide.
Primary apertures interiomarginal, umbilical, covered by a high tegillum.

Strat. distr. Ranging throughout *Globotruncana stuartiformis* zone, *Globotruncana gansseri* zone and *Globotruncana mayaroensis* zone.

Remarks See remarks *Globotruncana elevata* (Brotzen) and *Globotruncana stuartiformis* Dalbiez. Locality of figured specimen is sample L 1775, W. Irian.

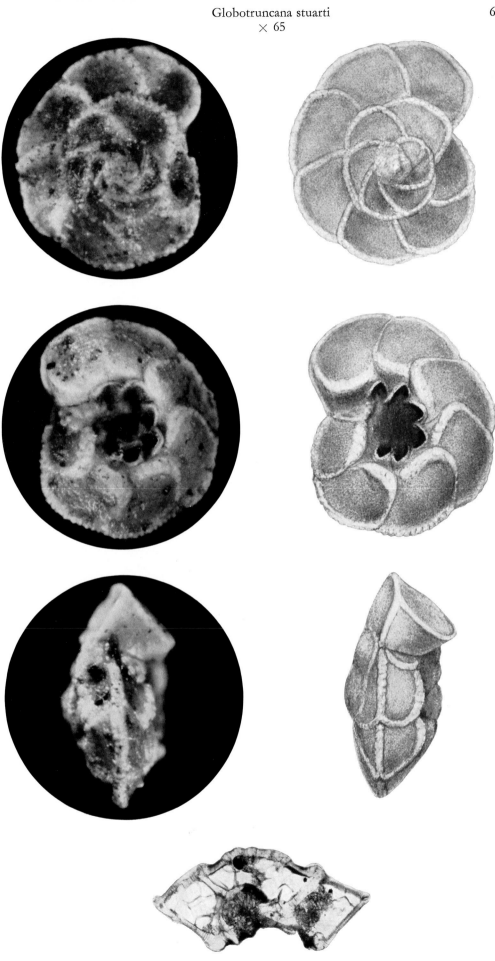

Globotruncana stuartiformis DALBIEZ

Reference *Globotruncana (Globotruncana) elevata stuartiformis* DALBIEZ, 1955: The genus *Globotruncana* in Tunisia. — Micropaleontology, 1 (2):169, text fig. 10a-c.

Type locality Le Kef-Mellegue, northwestern Tunisia.

Diagnosis Test very low trochospiral, central part of spiral side slightly convex, spiral side of last whorl almost flat, umbilical side convex; equatorial periphery slightly lobulate to almost circular; with one keel moderately beaded except in the last chamber.
Wall perforate, surface smooth.
Chambers subangular to angular, slightly inflated, often overlapping with a kind of carina on top of each chamber as a continuation of the raised sutures of the umbilical side; arranged in about 3 whorls, the 5-9 (usually 6-8) chambers of the last whorl increasing regularly in size.
Sutures on spiral side slightly curved in the first whorls to almost straight and tangential in the last whorl, raised, moderately beaded, on umbilical side curved, in first part of the last whorl raised and beaded, in later part flush.
Umbilicus deep, wide.
Primary apertures interiomarginal, umbilical, covered by a tegillum.

Strat. distr. Base *Globotruncana elevata* zone to top *Globotruncana stuartiformis* zone. Questionable occurrence in lower part of *Globotruncana gansseri* zone.

Remarks This species differs from *Globotruncana elevata* (BROTZEN) in the less convex umbilical side, the less distinct central cone on the spiral side, and the contour of the last-formed chambers, which are almost triangular instead of petaliform.
Globotruncana stuarti (DE LAPPARENT) has a more convex spiral side, no central cone, and the contour on the spiral side of the last-formed chambers is trapezoidal.
See also remarks *Globotruncana elevata* (BROTZEN).
Locality of figured specimen is Dyr el Kef section, sample 2F 165, W. Tunisia.

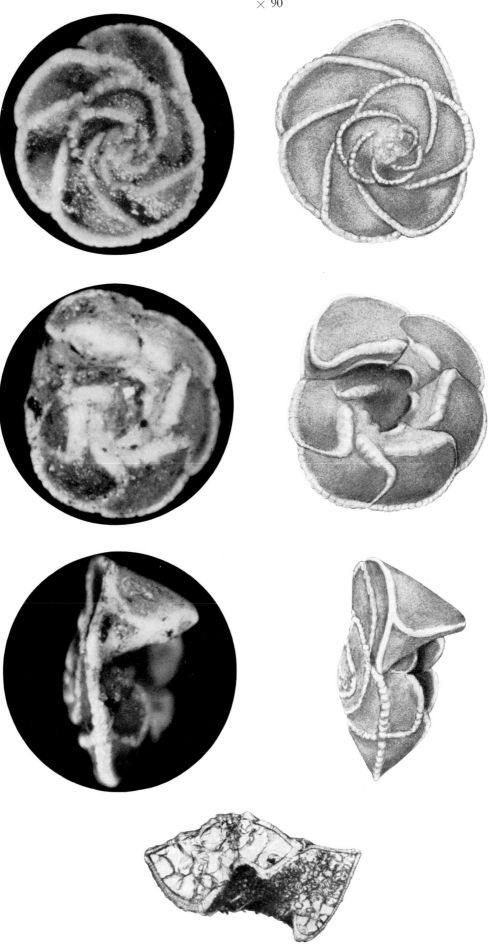

Globotruncana ventricosa WHITE

Reference *Globotruncana canaliculata* var. *ventricosa* WHITE, 1928: Some index Foraminifera of the Tampico embayment area of Mexico. — Journal of Paleontology, 2 (4):284, pl. 38, fig. 5a-c.

Type locality Two kilometres northeast of El Barranco on the road to Aldama, Tampico embayment, Mexico.

Diagnosis Test very low trochospiral, spiral side almost flat or slightly convex, umbilical side strongly convex; equatorial periphery lobulate with two fairly widely spaced keels, distinctly beaded, at least in the first portion.
Wall perforate, surface smooth.
Chambers angular, more or less inflated, each developing a carina on top, bordering the umbilicus; arranged in 2½-3 whorls, the 6-7 chambers increase moderately in size, showing on the spiral side a somewhat imbricate structure.
Sutures on spiral side curved, oblique, except the last ones, which become more radial, strongly raised, beaded; on umbilical side slightly curved, flush to depressed.
Umbilicus moderately deep, wide.
Primary apertures interiomarginal, umbilical, covered by a high tegillum.

Strat. distr. Lower part *Globotruncana elevata* zone to top *Globotruncana calcarata* zone. Questionable occurrence in lower part *Globotruncana stuartiformis* zone.

Remarks *Globotruncana ventricosa* might be related to *Globotruncana rosetta* (CARSEY) and *Globotruncana aegyptiaca* NAKKADY.
See remarks on *Globotruncana concavata* (BROTZEN) for differences between this species and *Globotruncana ventricosa* WHITE.
Locality of figured specimen is Dyr el Kef section, sample 2F187, W. Tunesia.

Globotruncana ventricosa
× 95

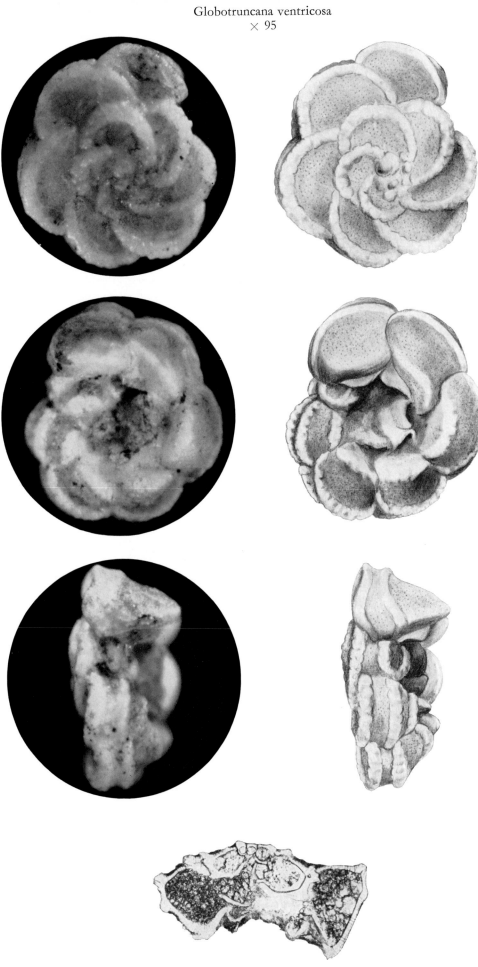

Hedbergella washitensis (Carsey)

Reference   *Globigerina washitensis* Carsey, 1926: Foraminifera of the Cretaceous of Central Texas. — The University of Texas Bulletin 2612:44, pl. 7, fig. 10.

Type locality   Shoal Creek at Austin, Travis County, Texas, U.S.A.

Diagnosis   Test low to medium trochospiral; equatorial periphery lobulate to distinctly lobulate. Wall perforate, surface with coarse reticulations, giving a honeycomb appearance, the elevated ridges leaving deep polygonal pits between them.
Chambers spherical, arranged in about $2\frac{1}{2}$ whorls, the 4-5 chambers of the last whorl increasing irregularly in size.
Sutures distinctly depressed, radial.
Aperture an interiomarginal, rather high arch, nearly umbilical in position, showing a slightly extra-umbilical tendency, bordered by a more or less distinct lip.

Strat. distr.   Base *Planomalina buxtorfi* zone to upper part *Rotalipora appenninica* zone. Questionable occurrence in upper part *Globigerinelloides breggiensis* zone and upper part *Rotalipora appenninica* zone.

Remarks   Locality of figured specimen is Dyr el Kef section, sample 1F 954, W. Tunisia.

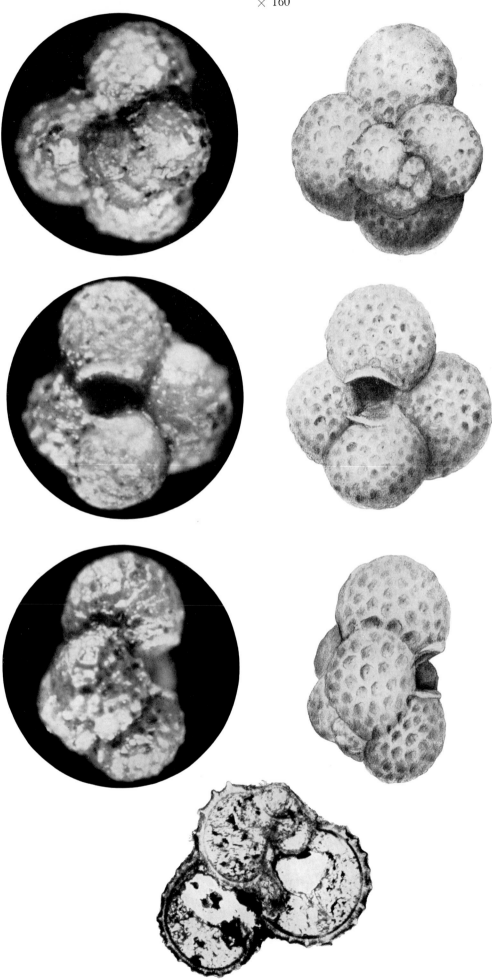

Planomalina buxtorfi (GANDOLFI)

Reference   *Planulina buxtorfi* GANDOLFI, 1942: Ricerche micropaleontologiche e stratigrafiche sulla Scaglia e sul Flysch cretacici dei dintorni di Balerna (Canton Ticino). — Rivista Italiana di Paleontologia, 48, Suppl. Mem. 4:103, pl. III, fig. 7.

Type locality   Gorge of the Breggia River, northeast of Balerna, near Chiasso, Canton of Ticino, southeastern Switzerland.

Diagnosis   Test planispiral, rather deeply biumbilicate, slightly evolute; equatorial periphery lobulate with a distinct keel.
Wall perforate, surface smooth.
Chambers more or less elongated angular-rhomboid; the 9-11 chambers of the last whorl increase regularly in size.
Sutures distinct, strongly curved and raised, early ones beaded to nodose.
Primary aperture interiomarginal, an equatorial arch with lateral extensions reaching back at either side to the septum at the base of the chamber and bordered by a distinct thickened lip, the lateral umbilical portions of successive apertures remaining visible as supplementary relict apertures.

Strat. distr.   Ranging throughout *Planomalina buxtorfi* zone.

Remarks   The genus *Planomalina* was established in 1946 by LOEBLICH and TAPPAN with the type species *Planomalina apsidostroba* LOEBLICH and TAPPAN, 1946, which, however, is a synonym of *Planomalina buxtorfi* (GANDOLFI), originally described in 1942 as *Planulina buxtorfi* GANDOLFI.
Most probably *Planomalina? almadenensis* CUSHMAN and TODD, 1948, is also a synonym of *Planomalina buxtorfi* (GANDOLFI).
Locality of figured specimen is Dyr el Kef section, sample 1F 961, W. Tunisia.

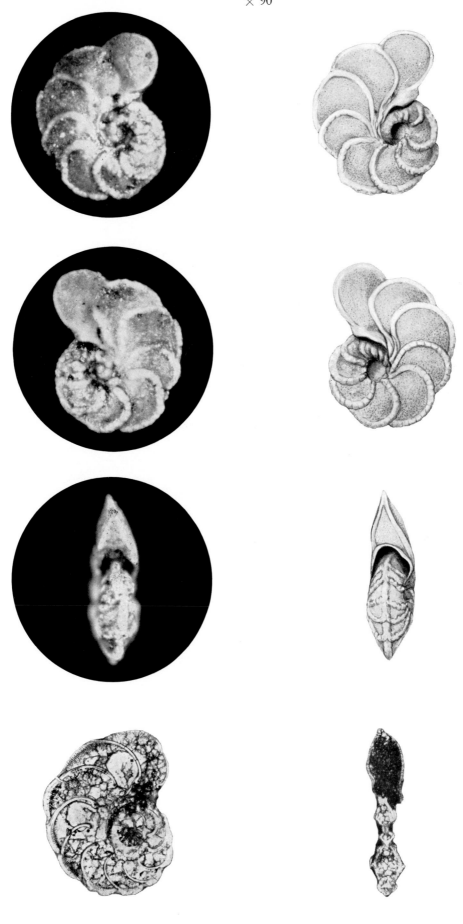

Praeglobotruncana citae (BOLLI)

Reference *Globotruncana citae* BOLLI, 1951: The genus *Globotruncana* in Trinidad, B.W.I. — Journal of Paleontology, 25 (2):197, pl. 35, figs. 4-6.

Type locality Outcrop in the river bed south of the bridge near mile post 12.5 on the Guaracara-Tabaquite Road, Lantern Estate, Central Range, Trinidad, B.W.I.

Diagnosis Test low trochospiral, biconvex to slightly spiroconvex; equatorial periphery moderately to distinctly lobulate, with a very moderate keel in the early chambers of the last whorl, which is hardly visible in the last chambers.
Wall perforate, surface mostly smooth; very fine spines may occasionally be present on the surface of the early chambers.
Chambers ovoid to subangular, arranged in 2½-3 whorls; the 4-5 chambers of the last whorl increase regularly in size, showing on the spiral side a slightly imbricate structure. Sutures depressed, on the spiral side curved, on the umbilical side straight and radial. Umbilicus shallow, fairly wide.
Aperture a relatively large, interiomarginal, extraumbilical-umbilical arch bordered by a prominent lip. The umbilical parts of the preceding lips may remain visible.

Strat. distr. Upper part *Globotruncana elevata* zone to top *Globotruncana mayaroensis* zone.

Remarks *Globotruncana havanensis* VOORWIJK, 1937, is closely related to *Praeglobotruncana citae* or even synonymous.
Locality of figured specimen is sample 1K 121, W. Tunisia.

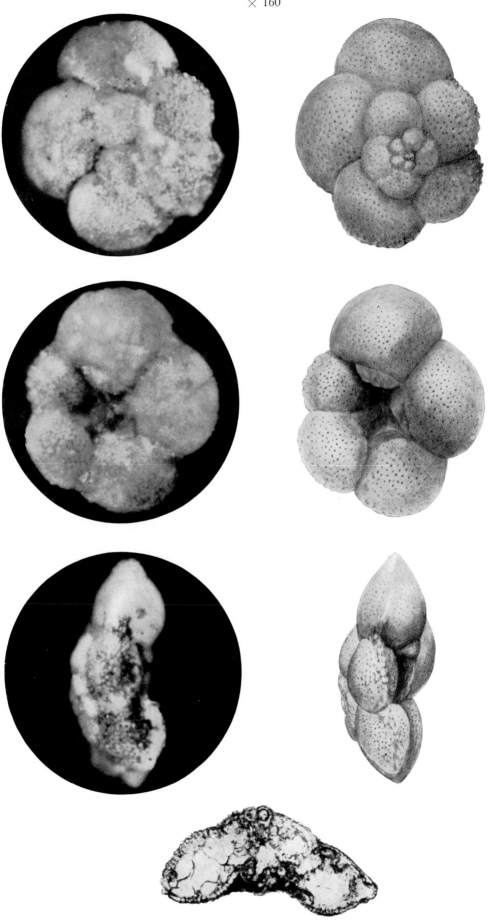

Praeglobotruncana stephani (GANDOLFI)

Reference   *Globotruncana stephani* GANDOLFI, 1942: Ricerche micropaleontologiche e stratigrafiche sulla Scaglia e sul Flysch cretacici dei dintorni di Balerna (Canton Ticino). — Rivista Italiana di Paleontologia, 48, Suppl. Mem. 4:130, pl. III, fig. 4.

Type locality   Gorge of the Breggia River, northeast of Balerna, near Chiasso Canton of Ticino, southeastern Switzerland.

Diagnosis   Test trochospiral, biconvex; equatorial periphery slightly lobulate with a nodose keel, which is smooth and less distinct in the last two or three chambers; the nodose keel of earlier whorls remains visible on the spiral side.
Wall perforate, surface of the first chambers of the last whorl moderately nodose, both on the umbilical side and on the spiral side.
Chambers subangular in the early part of the last whorl, the last ones being more rounded and inflated.
Sutures on spiral side gently curved, flush in the early whorls, depressed in the last whorl; on the umbilical side depressed, radial.
Umbilicus narrow, shallow.
Aperture a relatively low, interiomarginal, extraumbilical-umbilical arch, almost extending to the periphery, with a distinct bordering lip. The umbilical parts of the lips of the two or three preceding apertures remain visible.

Strat. distr.   Base *Rotalipora appenninica* zone into lower part *Rotalipora cushmani* zone. Questionable occurrence in upper part *Rotalipora cushmani* zone and lowermost part *Globotruncana helvetica* zone.

Remarks   It appears that *Praeglobotruncana stephani* and *Praeglobotruncana delrioensis* (PLUMMER) are very close variants. Several authors even regard these species as synonymous; LOEBLICH and TAPPAN (1961) are, on the other hand, of another opinion, and mention some differences.
Locality of figured specimen is Dyr el Kef section, sample 1F 961, W. Tunisia.

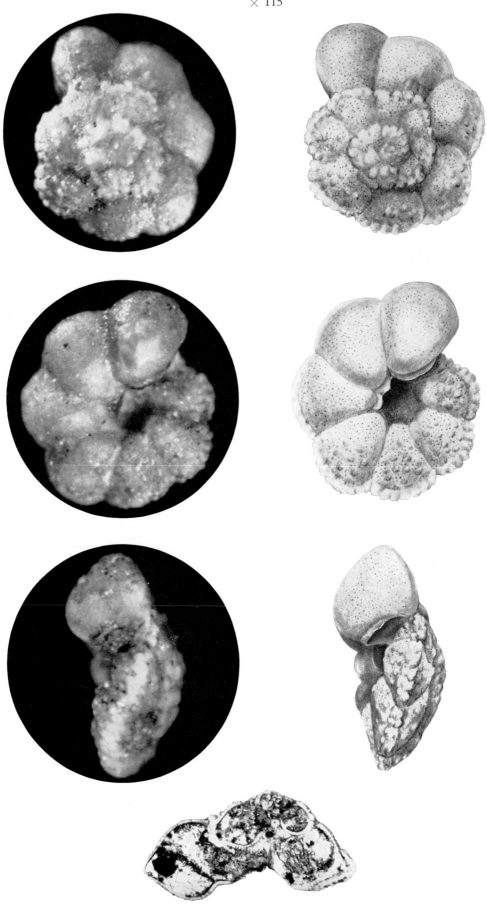

Praeglobotruncana turbinata (REICHEL)

Reference *Globotruncana stephani* GANDOLFI var. *turbinata* REICHEL, 1949: Observations sur les *Globotruncana* du gisement de la Breggia (Tessin). — Eclogae Geologicae Helvetiae, 42 (2):609.

Type locality Gorge of the Breggia River, northeast of Balerna, near Chiasso, Canton of Ticino, south-eastern Switzerland.

Diagnosis Test trochospiral, slightly biconvex to spiroconvex; equatorial periphery slightly lobulate with a nodose keel, which may be smooth and less distinct in the last two chambers; the nodose keel of earlier whorls remains visible on the spiral side.

Wall perforate, surface of the first chambers of the last whorl moderately nodose, both on the umbilical side, and on the spiral side.

Chambers subangular in the early part of the last whorl, the last ones being more inflated. Sutures on spiral side gently curved, raised and beaded, the last ones may be smooth; on the umbilical side depressed, radial.

Umbilicus fairly wide and deep.

Aperture a relatively low, interiomarginal, extraumbilical-umbilical arch, almost extending to the periphery, bordered by a flap, which may fuse with the preceding flap.

Strat. distr. Lower part *Rotalipora cushmani* zone into upper part *Globotruncana helvetica* zone. Questionable occurrence in uppermost part *Globotruncana helvetica* zone.

Remarks Locality of figured specimen is sample 3K 55, W. Tunisia.

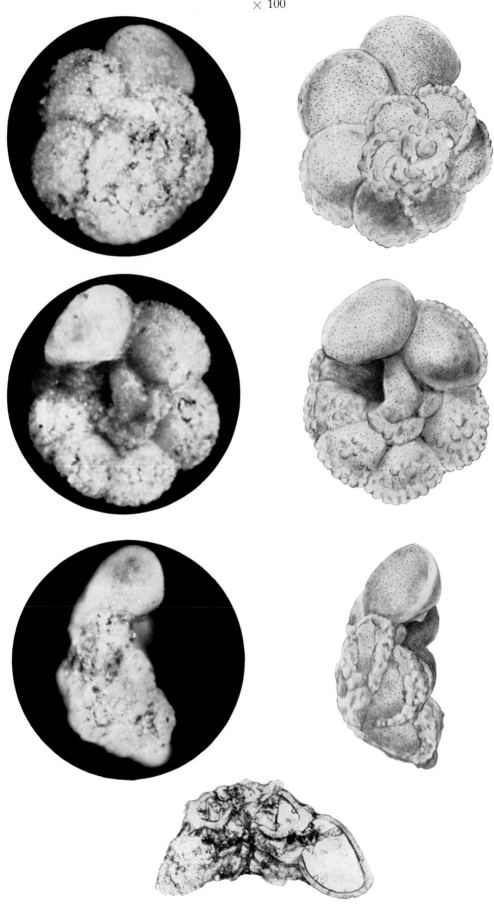

Rotalipora appenninica (Renz)

Reference *Globotruncana appenninica* Renz, 1936: Stratigrafische und micropalaeontologische Untersuchung der Scaglia (Obere Kreide-Tertiär) im Zentralen Apennin. — Eclogae Geologicae Helvetiae, 29 (1) : 20, text fig. 2.

Type locality Profile I, Bottaccione valley, northeastern Gubbio, Italy.

Diagnosis Test low trochospiral, biconvex; equatorial periphery lobulate, with one keel, which is weakly beaded in the early part.
Wall perforate, surface smooth.
Chambers angular-rhomboid, moderately inflated on the umbilical side; a smooth raised edge is present on top of the chambers on the umbilical side as continuation of the sutures; arranged in about 3 whorls, the 6-7 chambers of the last whorl increasing regularly in size.
Sutures on spiral side curved and oblique, raised, may be lightly beaded, especially in the first whorls; on umbilical side radial to gently curved, flush to raised in the early part of the last whorl, depressed in the later part.
Umbilicus fairly deep and narrow.
Primary aperture a fairly low, interiomarginal. extraumbilical-umbilical arch, bordered by a faint lip, which is only visible in the last chamber; single distinct sutural secondary apertures, bordered by a lip, climbing up to the position of the umbilical shoulder, visible in the last chambers only.

Strat. distr. Base *Rotalipora appenninica* zone into lowermost *Rotalipora cushmani* zone.

Remarks *Rotalipora appenninica* was originally described from thin sections, and at that time this species comprised practically all known single-keeled forms of the Cenomanian.
Gandolfi (1942) introduced a new species, *Globotruncana ticinensis,* and distinguished a number of varieties which were later given species rank:
Globotruncana appenninica var. α: *Rotalipora balernaensis* (Gandolfi)
,,   ,,   var. β: *Praeglobotruncana turbinata* (Reichel)
,,   ,,   var. γ: *Rotalipora reicheli* (Mornod)
,,   ,,   *s.s.* or *typica: Rotalipora appenninica* (Renz)
Locality of figured specimen is sample 1K 291, W. Tunisia.

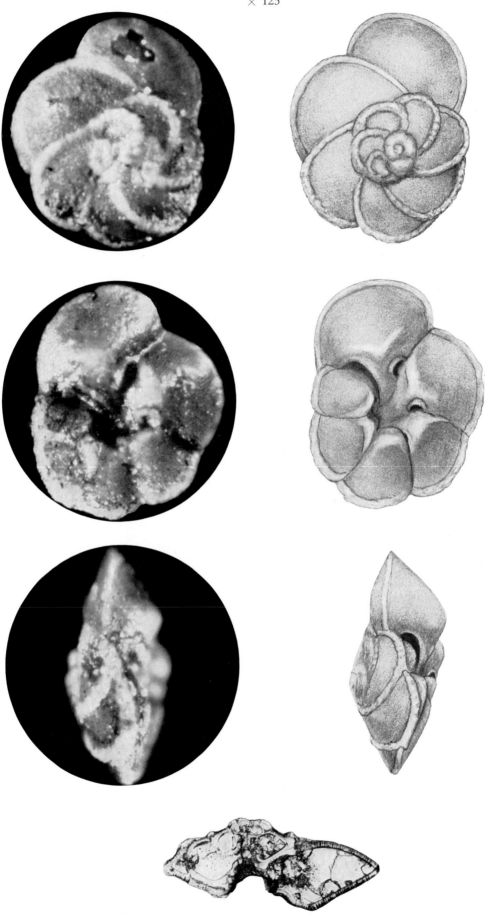

Rotalipora cushmani (MORROW)

Reference   *Globorotalia cushmani* MORROW, 1934: Foraminifera and Ostracoda from the Upper Cretaceous of Kansas. — Journal of Paleontology, 8 (2):199, pl. 31, figs. 2a-b, 4a-b.

Type locality   Sec. 31, T. 21S, R. 31W., Hodgeman County, Kansas, U.S.A.

Diagnosis   Test low trochospiral, biconvex; equatorial periphery lobulate to distinctly lobulate, with a weakly beaded narrow keel, which is mostly smooth in the last chambers.
Wall perforate, surface slightly rugose on both sides.
Chambers angular-rhomboid, strongly inflated on both sides; a three-sided thickened ridge may be present on top of the last chambers on the umbilical side, where the inflation is most prominent; arranged in about 2½ whorls, the 5-6 chambers of the last whorl increasing rather rapidly in size.
Sutures on spiral side curved, raised and weakly beaded in last whorl, depressed in initial whorls; on umbilical side radial, depressed.
Umbilicus fairly wide and deep.
Primary aperture a high, interiomarginal, extraumbilical-umbilical arch, bordered above by a distinct lip, only visible in the last chamber; a single elongated, well developed sutural secondary aperture bordered by a lip is present on the umbilical shoulder of each chamber.

Strat. distr.   Ranging throughout *Rotalipora cushmani* zone.

Remarks   The following species are considered synonyms:
        Rotalipora turonica BROTZEN, 1942
        Globotruncana alpina BOLLI, 1945
        Globotruncana (Rotalipora) montsalventis MORNOD, 1949
Locality of figured specimen is Neuweidgraben, sample 1,395m, Ammergau Mountains, southern Germany.

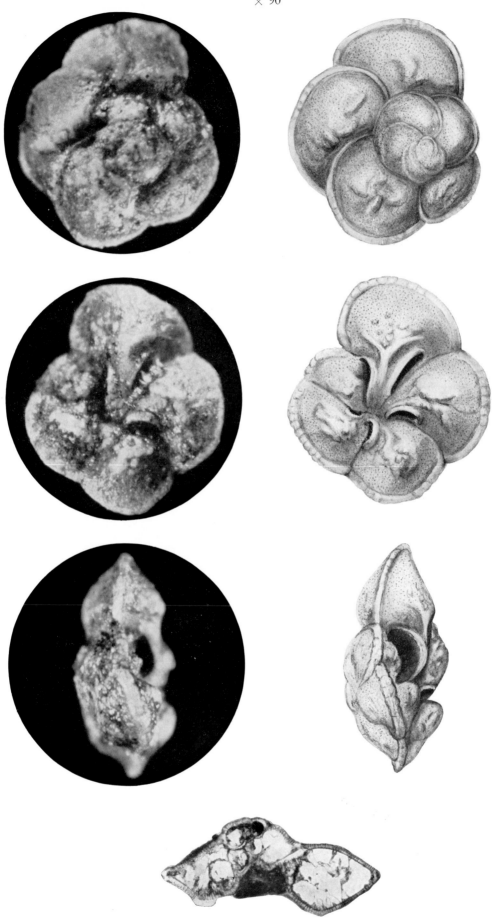

Rotalipora greenhornensis (MORROW)

Reference   *Globorotalia greenhornensis* MORROW, 1934: Foraminifera and Ostracoda from the Upper Cretaceous of Kansas. — Journal of Paleontology, 8:199, pl. 31, fig. 1.

Type locality   Sec. 31, T. 21S., R. 22W., Hodgeman County, Kansas, U.S.A.

Diagnosis   Test low trochospiral, biconvex; equatorial periphery slightly lobulate to almost circular with a very lightly beaded to smooth keel.

Wall perforate, surface usually smooth but may be somewhat nodose in the first chambers of the last whorl.

Chambers angular-rhomboid, arranged in 2½-3 whorls; the 6-8 chambers of the last whorl increase regularly and slowly in size.

Sutures on spiral side curved to slightly curved, oblique, but may become more radial between the last chambers, strongly raised, lightly beaded; on umbilical side curved, marked by lightly beaded septal carinae with periumbilical lengthenings, the septal carinae may be absent between the last two or three chambers.

Umbilicus deep, rather narrow to fairly wide.

Primary aperture a relatively high interiomarginal, extraumbilical-umbilical arch, bordered by a lip as part of an imperforate flange projecting from the chamber wall into the umbilicus which fuses with the flanges of the preceding chambers; a single sutural secondary aperture bordered by a rim is visible near the umbilical shoulder of most chambers.

Strat. distr.   Base *Rotalipora greenhornensis* zone to upper part *Rotalipora cushmani* zone.

Remarks   *Thalmanninella brotzeni* SIGAL and *Rotalipora globotruncanoides* SIGAL, both described in 1948 from Algerian material, are considered to be synonyms of *Rotalipora greenhornensis*.

Locality of figured specimen is Dyr el Kef section, sample 2F 10, W. Tunisia.

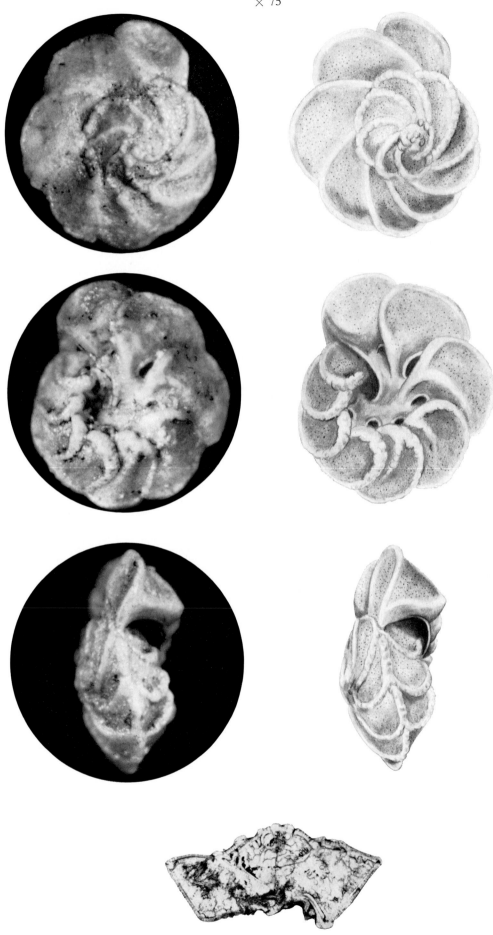

Rotalipora reicheli (MORNOD)

Reference *Globotruncana (Rotalipora) reicheli* MORNOD, 1949: Les Globorotalidés du Crétacé supérieur du Montsalvens (Préalpes fribourgeoises). — Eclogae Geologicae Helvetiae, 42 (2):583, text fig. 5, IVa-c.

Type locality Upper part of Profile III, at about 985 metres above sea level, in the Ruisseau des Covayes, on the southeastern slope of the Montsalvens chain, north of Cerniat, in the Préalpes fribourgeoises, Switzerland.

Diagnosis Test low trochospiral, central part of spiral side convex, spiral side of last whorl flat to concave, umbilical side strongly convex, especially the last chambers; equatorial periphery slightly lobulate with a beaded keel, which becomes narrow and smooth in the last chambers.

Wall perforate; surface smooth, except the umbilical shoulders in the later portion of the test, which may be rugose.

Chambers angular to subangular, last ones becoming strongly inflated and steep on umbilical side; a nodose raised edge is present on top of the chambers on the umbilical side as continuation of the sutures; arranged in about 3 whorls, the 6-8 chambers of the last whorl sometimes increasing irregularly in size.

Sutures on spiral side oblique and curved, raised, distinctly beaded, especially in the first whorls; on umbilical side gently curved to radial, raised in the early portion of the last whorl, depressed in the later portion.

Umbilicus wide and fairly deep.

Primary aperture a high interiomarginal, extraumbilical-umbilical arch, bordered by a lip as part of an imperforate flange projecting from the chamber wall into the umbilicus which fuses with the flanges of the preceding chambers; a single distinct sutural secondary aperture bordered by a rim is present near the umbilical shoulder of each chamber.

Strat. distr. Ranging throughout upper part *Rotalipora cushmani* zone, extending a little into the lower part of this zone.

Remarks Locality of figured specimen is Dyr el Kef section, sample 1F 601, W. Tunisia.

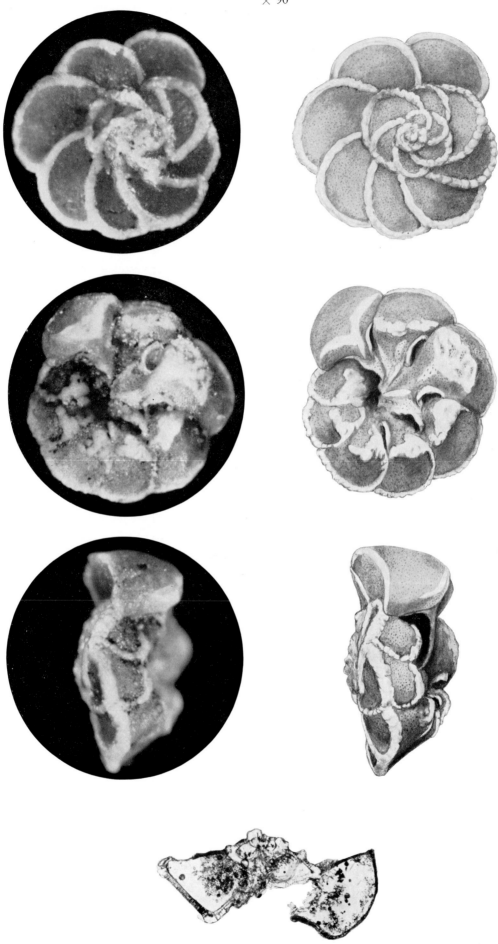

Rotalipora subticinensis (GANDOLFI)

Reference
Globotruncana ticinensis GANDOLFI var. α, 1942: Ricerche micropaleontologiche e stratigrafiche sulla Scaglia e sul Flysch cretacici dei dintorni di Balerna (Canton Ticino). — Rivista Italiana di Paleontologia, 48, Suppl. Mem. 4:114, pl. II, fig. 4
Globotruncana (Thalmanninella) ticinensis subticinensis GANDOLFI, 1957: Notes on some species of *Globotruncana*. — Contr. Cushman Found. Foram. Res., VIII:59, pl. 8, fig. 2.

Type locality
Gorge of the Breggia River, northeast of Balerna, near Chiasso, Canton of Ticino, southeastern Switzerland.

Diagnosis
Test low trochospiral, biconvex; equatorial periphery slightly lobulate to almost circular, with a faint keel, which is nodose in the early part.
Wall perforate; on the spiral side, surface of the chambers of the inner whorls and of the first three or four chambers of the last whorl distinctly nodose; on the umbilical side, surface of the first three or four chambers distinctly nodose.
Chambers angular in the penultimate whorl, subangular in the early part of the last whorl, the last ones becoming rounded and inflated.
Sutures on spiral side slightly curved, depressed, on umbilical side radial, depressed.
Umbilicus shallow, relatively narrow.
Primary aperture an interiomarginal, extraumbilical-umbilical arch bordered by a lip only visible in the last chamber; single sutural secondary apertures bordered by a faint rim, always situated on the umbilical side of and below the umbilical shoulders, are fairly distinct in the last chambers only.

Strat. distr.
Ranging throughout the *Rotalipora subticinensis* zone and the lower part of the *Globigerinelloides breggiensis* zone.

Remarks
Locality of figured specimen is Dyr el Kef section, sample 1F 936, W. Tunisia.

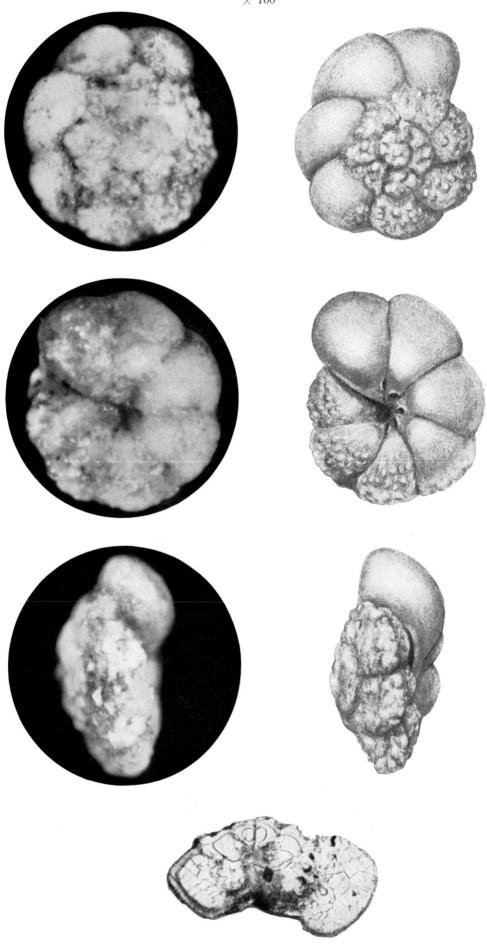

Rotalipora ticinensis (GANDOLFI)

Reference *Globotruncana ticinensis* GANDOLFI, 1942: Ricerche micropaleontologiche e stratigrafiche sulla Scaglia e sul Flysch cretacici dei dintorni di Balerna (Canton Ticino). — Rivista Italiana di Paleontologia, 48, Suppl. Mem. 4:113, pl. II, fig. 3.

Type locality Gorge of the Breggia River, northeast of Balerna, near Chiasso, Canton of Ticino, southeastern Switzerland.

Diagnosis Test low trochospiral, biconvex; equatorial periphery slightly lobulate to almost circular, with a beaded keel which may be smooth in the last chamber.
Wall perforate, surface smooth.
Chambers angular, rather compressed; arranged in 2½-3 whorls, the 7-8 chambers of the last whorl increasing slowly in size.
Sutures on spiral side curved, moderately oblique, raised, lightly beaded; on umbilical side radial to gently curved, raised to flush in the early part of the last whorl, depressed in the later part.
Umbilicus fairly wide and shallow.
Primary aperture a fairly high, interiomarginal, extraumbilical-umbilical arch bordered by a lip, only visible in the last chamber; single sutural secondary apertures, bordered by a rim, always situated under the umbilical shoulders, are distinct in the last chambers only.

Strat. distr. Base *Globigerinelloides breggiensis* zone into lower part *Rotalipora appenninica* zone.

Remarks Locality of figured specimen is sample G 731, W. Tunisia.

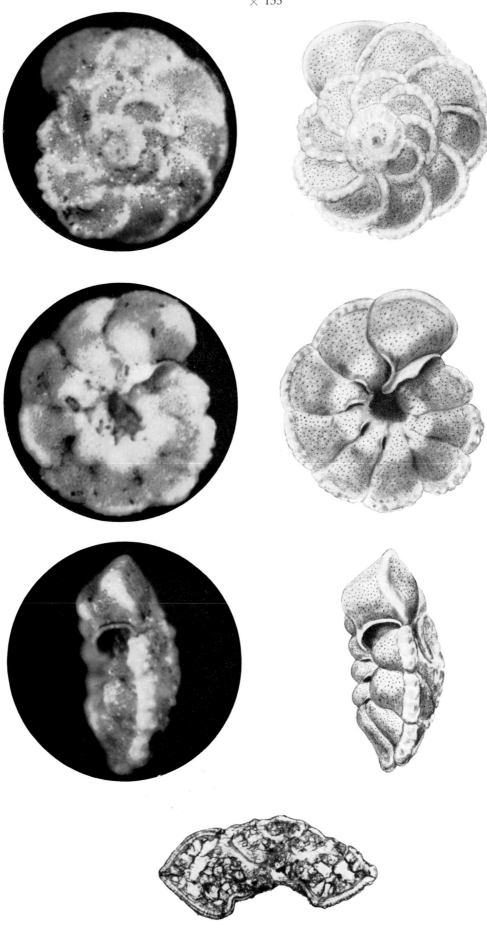

Rugoglobigerina rotundata BRONNIMANN

Reference   *Rugoglobigerina rugosa rotundata* BONNIMANN, 1952: Globigerinidae from the Upper Cretaceous (Cenomanian-Maestrichtian) of Trinidad, B.W.I. — Bulletin of American Paleontology, 34 (140):34, pl. 4, figs. 7-9.

Type locality   Trinidad Leaseholds Ltd. Catalogue nos. 155591-155594, subsurface samples from the Guayaguayare area, southeastern Trinidad, B.W.I.

Diagnosis   Test starts low trochospiral, followed in the adult by a higher whorl; equatorial periphery lobulate.
Wall perforate, surface rugose with numerous densely placed pustules, decreasing in size towards the last chamber.
Chambers spherical, truncate towards the apertures; arranged in about $2\frac{1}{2}$ whorls, the 5-6 chambers of the last whorl increasing moderately in size.
Sutures depressed, on the spiral side straight to slightly curved, on the umbilical side straight.
Umbilicus fairly wide, deep.
Primary apertures interiomarginal, umbilical, covered by a tegillum.

Strat. distr.   Ranging throughout *Globotruncana gansseri* zone and *Globotruncana mayaroensis* zone.

Remarks   Locality of figured specimen is sample 1K 117, W. Tunisia.

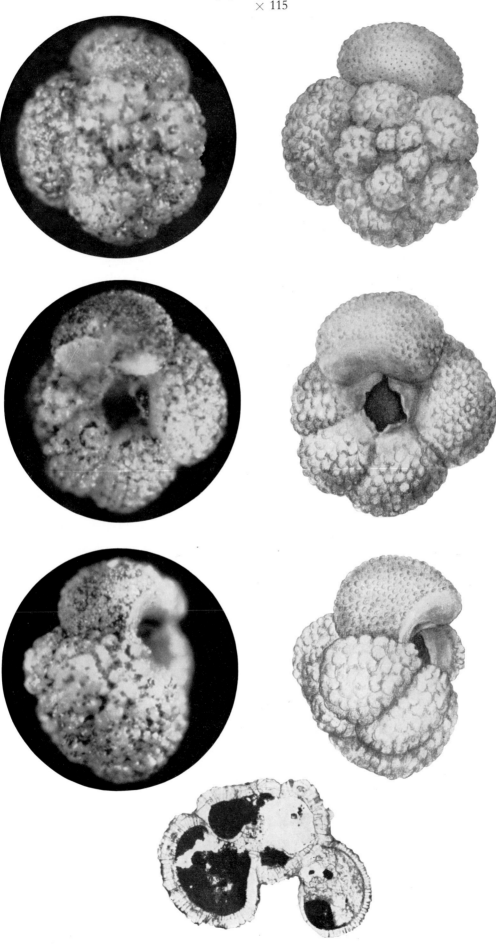

Rugoglobigerina rugosa (Plummer)

Reference *Globigerina rugosa* Plummer, 1926: Foraminifera of the Midway formation in Texas. — The University of Texas Bulletin, 2644:38, pl. II, fig. 10a-d.

Type locality From bank of Walker Creek, 6 miles N. 15° E. of Cameron, Milam County, Texas, U.S.A.

Diagnosis Test low trochospiral, equatorial periphery distinctly lobulate.
Wall perforate, surface of the chambers of the last whorl rugose with numerous large pustules which often coalesce into distinct ridges, radiating from the midpoint of each chamber on the periphery.
Chambers spherical, truncate towards the apertures; arranged in about $2\frac{1}{2}$ whorls, the 4-6 chambers of the last whorl increasing rapidly in size.
Sutures depressed, on the spiral side straight to slightly curved, on the umbilical side straight.
Umbilicus wide, fairly deep.
Primary apertures interiomarginal, umbilical, covered by a tegillum.

Strat. distr. Upper part *Globotruncana calcarata* zone to top *Globotruncana mayaroensis* zone. Questionable occurrence in upper part *Globotruncana elevata* zone and lower part *Globotruncana calcarata* zone.

Remarks Locality of figured specimen is Well Guayaguayare 163, core sample 5588'-5598', Trinidad, B.W.I.

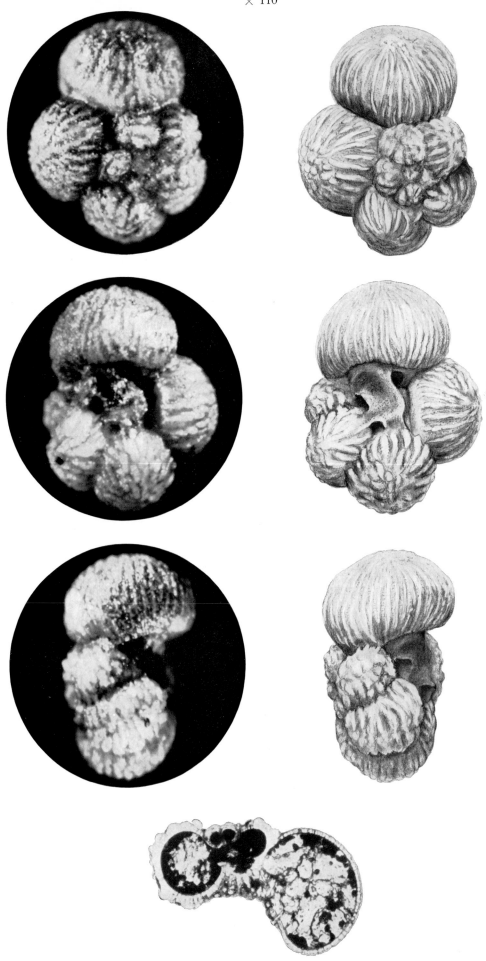

Rugoglobigerina scotti (Bronnimann)

Reference *Trinitella scotti* Bronnimann, 1952: Globigerinidae from the Upper Cretaceous (Cenomanian-Maestrichtian) of Trinidad, B.W.I. — Bulletin of American Paleontology, 34 (140): 57, pl. 4, figs. 4-6.

Type locality Trinidad Leaseholds Ltd. Catalogue nos. 155591-155594, subsurface samples from the Guayaguayare area, southeastern Trinidad, B.W.I.

Diagnosis Test very low trochospiral, spiral side almost flat, umbilical side moderately convex; equatorial periphery lobulate.

Wall perforate, surface of the chambers strongly rugose (pustules and ridges), except the last one, which is mostly smooth.

Chambers of the initial whorls and the first chambers of the last whorl are subglobular, gradually increasing in size, the last chambers becoming flattened on the spiral side and abruptly enlarged, about twice as large as the penultimate ones; arranged in $2\frac{1}{2}$-3 whorls, the last whorl consisting of 5-6 chambers.

Sutures depressed, on the spiral side strongly curved, on the umbilical side relatively straight to slightly curved. Some specimens show elevated sutures between the last chambers on the spiral side, in which case an indistinct pseudo-keel may be present. Umbilicus fairly wide.

Primary apertures interiomarginal, umbilical, covered by a delicate tegillum.

Strat. distr. Ranging throughout *Globotruncana mayaroensis* zone.

Remarks Locality of figured specimen is Dyr el Kef section, sample 2F 228, W. Tunisia.

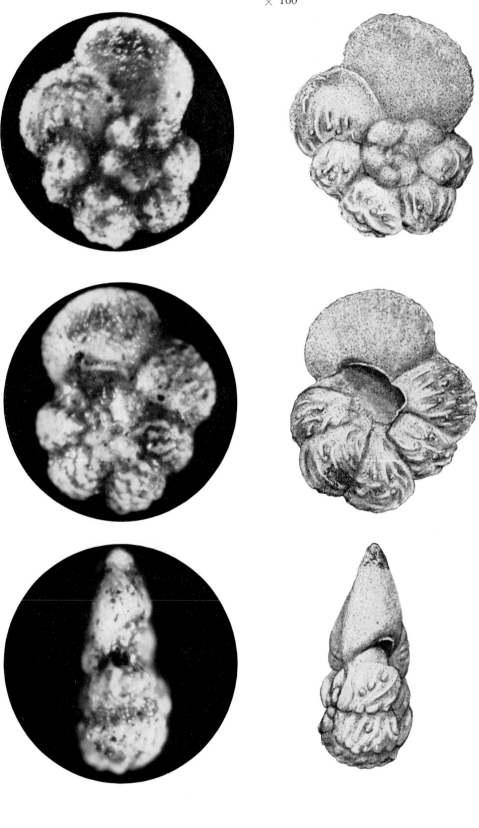

Ticinella roberti (Gandolfi)

Reference *Anomalina roberti* Gandolfi, 1942: Ricerche micropaleontologiche e stratigrafiche sulla Scaglia e sul Flysch cretacici dei dintorni di Balerna (Canton Ticino). — Rivista Italiana di Paleontologia, 48, Suppl. Mem. 4:100, pl. II, fig. 2.

Type locality Gorge of the Breggia River, northeast of Balerna, near Chiasso, Canton of Ticino, south-eastern Switzerland.

Diagnosis Test low trochospiral, the last evolution may be in a higher whorl; equatorial periphery lobulate.

Wall perforate, surface of the early chambers somewhat rugose, later chambers of the last whorl smooth.

Chambers subglobular, especially somewhat flattened on the spiral side, arranged in $2\frac{1}{2}$-3 whorls, the usually 8 chambers of the last whorl increasing gradually in size.

Sutures depressed, gently curved on the spiral side, nearly straight and radial on the umbilical side.

Umbilicus fairly wide.

Primary aperture a low, interiomarginal, extraumbilical-umbilical arch bordered by a lip as part of an imperforate flange projecting from the chamber wall into the umbilicus, fusing with the flanges of the preceding chambers; sutural secondary apertures bordered by rims are present near the umbilical margin.

Strat. distr. Ranging throughout *Ticinella roberti* zone to top *Planomalina buxtorfi* zone. Questionable occurrence in lower part *Rotalipora appenninica* zone.

Remarks Locality of figured specimen is Dyr el Kef section, sample 1F 932, W. Tunisia.

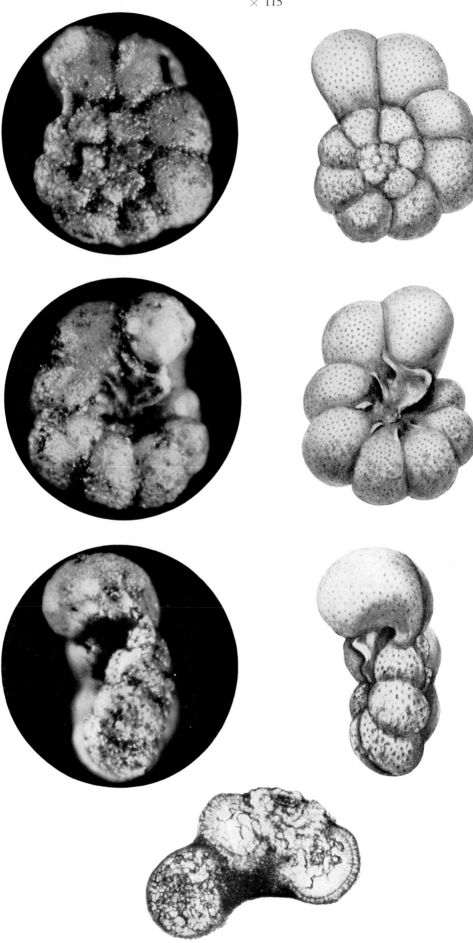

MESOZOIC ASSEMBLAGES IN THIN SECTIONS OF HARD ROCKS

Figure 1 — Foraminiferal wackestone with *Ticinella roberti* (GANDOLFI) and *"Ticinella" lorneiana* (D'ORRBIGNY). $\times$ 25
Sample Bn 103 of a section between Probbico and San Lorenzo, Marches-Umbria area, Italy.
Ticinella roberti zone.

Figure 2 — Foraminiferal wackestone with *Rotalipora subticinensis* (GANDOLFI). $\times$ 17
Sample Bn 306 of the Monte Torre section, Marches-Umbria area, Italy.
Rotalipora subticinensis zone.

Figure 3 — Foraminiferal wackestone with *Globigerinelloides breggiensis* (GANDOLFI) and *Rotalipora ticinensis* (GANDOLFI). $\times$ 20
Sample Bn 307 of the Monte Torre section, Marches-Umbria area, Italy.
Globigerinelloides breggiensis zone.

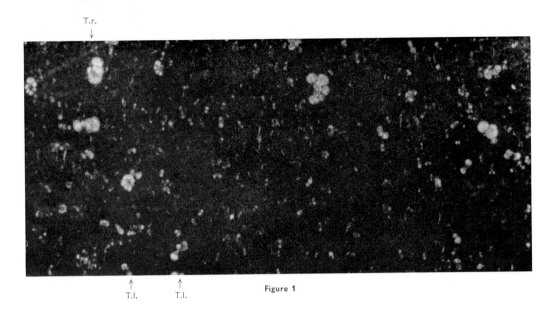

Figure 1

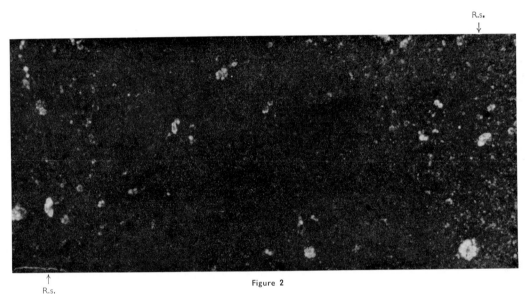

Figure 2

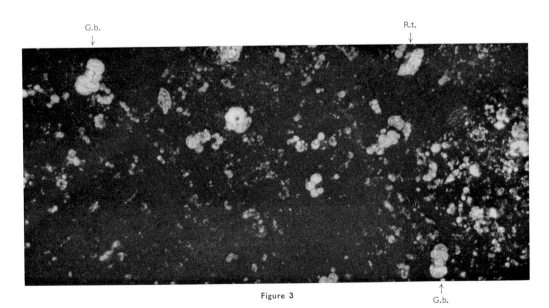

Figure 3

Figure 4 — Foraminiferal wackestone with *Planomalina buxtorfi* (GANDOLFI). × 22.5
Sample Bn 309 of the Monte Torre section, Marches-Umbria area, Italy.
Planomalina buxtorfi zone.

Figure 5 — Foraminiferal wackestone with *Rotalipora appenninica* (RENZ) and *Prae-
globotruncana stephani* (GANDOLFI). × 18
Sample Bn 252 of the Gubbio section, Marches-Umbria area, Italy (near type locality of
Rotalipora appenninica).
Rotalipora appenninica zone.

Figure 6 — Foraminiferal wackestone with *Rotalipora cushmani* (MORROW). × 17
Sample Ni 799 of the Murree Brewery section, near Quetta, W. Pakistan.
Rotalipora cushmani zone.

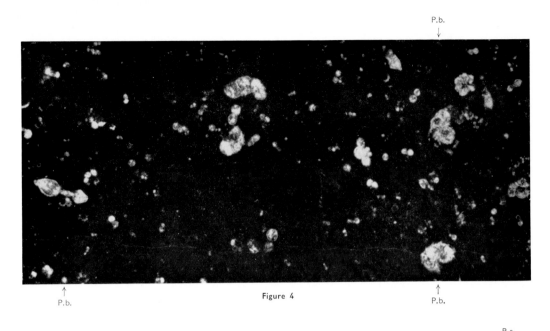

Figure 4

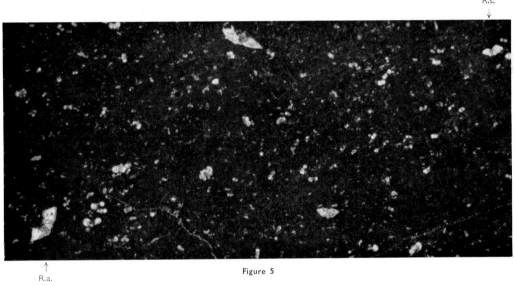

Figure 5

Figure 6

Figure 7 — Foraminiferal wackestone with *Globotruncana helvetica* BOLLI, *Globotruncana imbricata* MORNOD and *Praeglobotruncana turbinata* (REICHEL). ⨉ 18
Sample Sc. 1010 of the Peromanda section, Loralai area, W. Pakistan.
Globotruncana helvetica zone.

Figure 8 — Foraminiferal wackestone with *Globotruncana imbricata* MORNOD and double-keeled globotruncanas. ⨉ 20
Sample Sc. 1015 of the Peromanda section, Loralai area, W. Pakistan.
Globotruncana schneegansi zone.

Figure 9 — Foraminiferal wackestone with *Globotruncana concavata* (BROTZEN). ⨉ 18
Sample Sc. 1312 of the Dilkuna section, Loralai area, W. Pakistan.
Globotruncana concavata zone.

P.t.

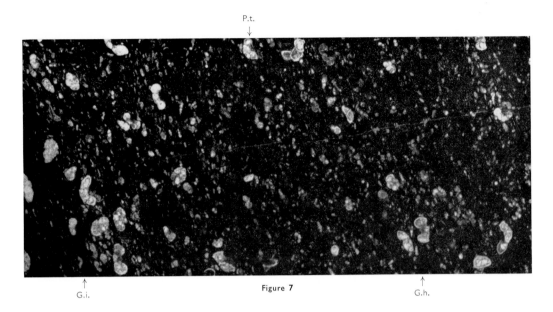

G.i.   Figure 7   G.h.

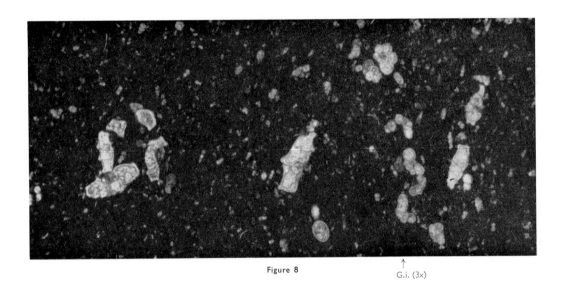

Figure 8   G.i. (3x)

G.c.

G.c.   Figure 9

Figure 10 — Foraminiferal wackestone with *Globotruncana carinata* DALBIEZ.   × 20
Sample Sc. 1417 of the Sembar section, Loralai area, W. Pakistan.
Globotruncana carinata zone.

Figure 11 — Foraminiferal wackestone with *Globotruncana elevata* (BROTZEN).   × 17
Sample Sc. 474 of the Tabela Char section, Loralai area, W. Pakistan.
Globotruncana elevata zone.

Figure 12 — Foraminiferal packstone with *Globotruncana contusa* (CUSHMAN), *Pseudo-textularia elegans* (RZEHAK) and *Globotruncana stuarti* (DE LAPPARENT).   × 18
Sample Bn 383 of the Fossombrone section, Marches-Umbria area, Italy.
Globotruncana gansseri zone – *Globotruncana mayaroensis* zone.

G. ca. (2x)

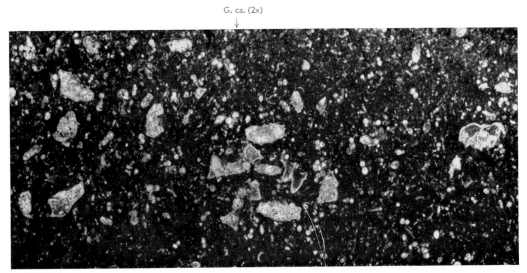

Figure 10

G.e.

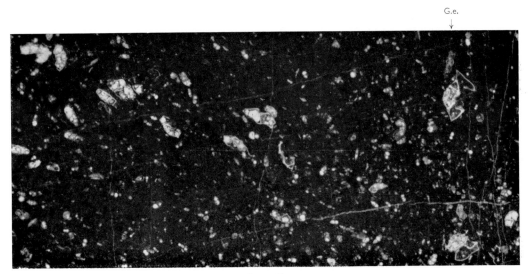

Figure 11

G.co.

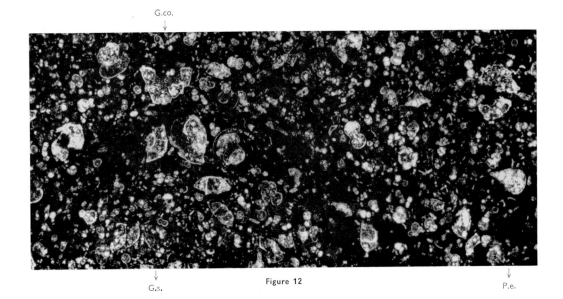

G.s.

Figure 12

P.e.

II. CENOZOIC

KEY TO THE CENOZOIC GENERA

I   Test planispiral throughout

   A   Aperture equatorial

      1   chambers spherical to ovate (early stage may be slightly trochospiral)
         — *Hastigerina* THOMSON, 1876

      2   chambers spherical in early stage, in adult radially elongate or clavate
         — *Clavigerinella* BOLLI, LOEBLICH and TAPPAN, 1957

      3   chambers subglobular or radially elongate with tubulospines
         — *Hantkenina* CUSHMAN, 1924

   B   Primary aperture equatorial in position with secondary multiple areal apertures consisting of small rounded openings above the primary aperture
      chambers subglobular with tubulospines
         — *Cribrohantkenina* THALMANN, 1942

II   Test planispiral in early stage, becoming enrolled biserial in adult

   Aperture in young stage equatorial, in adult extraumbilical
      chambers globular to ovate
         — *Cassigerinella* POKORNY, 1955

III   Test trochospiral throughout

   A   Aperture umbilical

      1   without bulla(e)

         a.   aperture with or without a thin lip, no secondary apertures present
            — *Globigerina* D'ORBIGNY, 1826

         b.   aperture covered above by a flap or umbilical tooth, no secondary apertures present
            — *Globoquadrina* FINLAY, 1947

         c.   aperture (primary) with or without a thin lip, with sutural secondary apertures   — *Globigerinoides* CUSHMAN, 1927

         d.   aperture in adult replaced by multiple sutural secondary apertures
            — *Candeina* D'ORBIGNY, 1839

      2   with bulla(e)

         a.   primary aperture in final stage covered by a single umbilical bulla, with one or more infralaminal secondary apertures (no sutural secondary apertures)   — *Catapsydrax* BOLLI, LOEBLICH and TAPPAN, 1957

         b.   primary aperture in final stage covered by an irregular bulla expanding along the earlier sutures, with numerous infralaminal secondary apertures (no sutural secondary apertures)
            — *Globigerinita* BRONNIMANN, 1951

         c.   primary aperture in final stage covered by an umbilical bulla, one or more sutural secondary apertures are present, which are usually covered by sutural bullae, with infralaminal secondary apertures
            — *Globigerinoita* BRONNIMANN, 1952

B Aperture extraumbilical-umbilical

without bulla(e)

a. without sutural secondary apertures

   i. chambers ovate to angular rhomboid or angular conical, with or without a single keel
       — *Globorotalia* CUSHMAN, 1927

   ii. adult chambers radially elongate, clavate or cylindrical-no keel
       — *Hastigerinella* CUSHMAN, 1927

b. with sutural secondary apertures on the spiral side
    — *Truncorotaloides* BRONNIMANN and BERMUDEZ, 1953

IV Test trochospiral in early stage, last chamber(s) embracing partly or completely the trochospiral stage which shows an umbilical aperture

1 without bulla(e)

a. early stage *Globigerina*-like, final chamber, provided with sutural secondary apertures, embracing umbilical region
    — *Globigerapsis* BOLLI, LOEBLICH and TAPPAN, 1957

b. early stage *Globigerinoides*-like, final chamber, provided with sutural secondary apertures, embracing umbilical region
    — *Orbulinoides* BLOW and SAITO, 1968 or
    — *Praeorbulina* OLSSON, 1964

c. initial stage *Globigerina*-like, final chamber embracing nearly or completely initial stage; aperture areal over a great part of the test, small sutural secondary apertures around the early Globigerine chambers where these are visible   — *Orbulina* D'ORBIGNY, 1839

2 with bulla(e)

a. early stage *Globigerina*-like, enveloping final chamber covering previous umbilical side, with sutural secondary apertures, which are covered by small bullae, each with one or more infralaminal openings
    — *Globigerinatheka* BRONNIMANN, 1952

b. like *Globigerinatheka*, but with the sutural bullae more irregularly distributed and later chambers provided with secondary areal apertures covered by small knobby pustule-like bullae; all bullae may have infralaminal openings
    — *Globigerinatella* CUSHMAN and STAINFORTH, 1945

3 with or without bulla(e)

early stage *Globigerina*-like, in final stage two or three embracing chambers with one or more sutural secondary apertures, which may be obscured by overhanging chamber flanges. Wall structure in later chambers complex, consisting of more than one layer of shell material—small bulla(e) may occasionally be present
    — *Sphaeroidinella* CUSHMAN, 1927

V Test trochospiral in early stage, becoming streptospiral in adult. In young stage *Globigerina*-like with open umbilicus, in later stage no umbilicus is present as the axis of coiling changes
    — *Pulleniatina* CUSHMAN, 1927

DESCRIPTIONS OF THE CENOZOIC GENERA

Genus *Candeina* D'Orbigny, 1839

Reference   Orbigny, A.D. D', 1839: Foraminifères *in* de la Sagra, Histoire physique, politique et naturelle de l'Ile de Cuba. — p. 107.

Type species   *Candeina nitida* D'Orbigny, 1839.

Diagnosis   Test trochospiral, relatively high-spired.
Wall finely perforate, surface smooth.
Chambers globular to hemispherical.
Sutures radial to curved, depressed.
Primary aperture in the early stage interiomarginal, umbilical, later in development smaller secondary sutural apertures occur on each side of the primary aperture; in the adult tests there is no primary opening and the small rounded sutural secondary apertures almost completely surround the later chambers.

Genus *Cassigerinella* Pokorny, 1955

Reference   Pokorny, V., 1955: *Cassigerinella boudecensis* n.gen., n.sp. (Foraminifera, Protozoa) z oligocénu ždánického flyše. — Věstník Ústředního Ústavu Geologického, 30:136.

Type species   *Cassigerinella boudecensis* Pokorny, 1955.

Diagnosis   Test in early stage planispiral, later with biserially arranged chambers continuing to spiral in the same plane, biumbilicate.
Wall perforate, surface smooth to pitted.
Chambers globular to ovate.
Sutures radial to curved, depressed.
Aperture an interiomarginal, extraumbilical arch, alternating in position from one side to the next in successive chambers.

Genus *Catapsydrax* Bolli, Loeblich and Tappan, 1957

Reference   Bolli, H.M., Loeblich, A.R. and Tappan, H., 1957: Planktonic Foraminiferal Families Hantkeninidae, Orbulinidae, Globorotaliidae and Globotruncanidae. — United States National Museum Bulletin, 215:36.

Type species   *Globigerina dissimilis* Cushman and Bermudez, 1937.

Diagnosis   Test trochospiral.
Wall perforate, surface smooth or pitted.
Chambers spherical to ovate.
Sutures radial to slightly curved, depressed.
Primary aperture interiomarginal, umbilical, in the final stage covered by a single umbilical bulla, with one or more accessory infralaminal apertures.

Genus *Clavigerinella* Bolli, Loeblich and Tappan, 1957

Reference   Bolli, H.M., Loeblich, A.R. and Tappan, H., 1957: Planktonic Foraminiferal Families Hantkeninidae, Orbulinidae, Globorotaliidae and Globotruncanidae. — United States National Museum Bulletin, 215:30.

Type species   *Clavigerinella akersi* Bolli, Loeblich and Tappan, 1957.

Diagnosis   Test planispiral, biumbilicate, involute; radially lobulate in outline.
Wall perforate, surface finely pitted.
Chambers of the first whorls spherical, chambers of the last whorl radially elongate or clavate.
Sutures radial, depressed.
Aperture interiomarginal, equatorial, an elongate slit extending up the apertural face, bordered laterally by wide flanges.

Genus *Cribrohantkenina* THALMANN, 1942

Reference THALMANN, H.E., 1942: Foraminiferal genus *Hantkenina* and its subgenera. — American Journal of Science, 240:812.

Type species *Hantkenina (Cribrohantkenina) bermudezi* THALMANN, 1942.

Diagnosis Test planispiral, biumbilicate, almost involute, biconvex.
Wall perforate, surface smooth.
Chambers subglobular, inflated, with a prominent peripheral tubulospine at the forward margin of each chamber, the succeeding chambers being attached near the base of the tubulospines; chambers may partially or completely envelop the tubulospine of the preceding chamber.
Sutures almost radial, depressed.
Primary aperture an interiomarginal, equatorial arch, projecting into the umbilical regions, bordered at top by a distinct lip; secondary apertures are areal supplementary apertures (two or more) arranged in one or more rows above the primary 'aperture, bordered by distinct protruding lips.

Genus *Globigerapsis* BOLLI, LOEBLICH and TAPPAN, 1957

Reference BOLLI, H.M., LOEBLICH, A.R. and TAPPAN, H., 1957: Planktonic foraminiferal families Hantkeninidae, Orbulinidae, Globorotaliidae and Globotruncanidae. — United States National Museum Bulletin, 215:33.

Type species *Globigerapsis kugleri* BOLLI, LOEBLICH and TAPPAN, 1957.

Diagnosis Test subglobular, early portion trochospiral with subglobular chambers, final chamber embracing and covering the umbilical region of the early coil.
Wall perforate, surface smooth to hispid or spinose.
Sutures radial to curved, depressed.
Primary aperture interiomarginal, umbilical in the young stage, covered in the adult by an enveloping final chamber, with two or more arched secondary apertures at the lower margin of the final chamber, at the contact with the sutures of the earlier whorl.

Genus *Globigerina* D'ORBIGNY, 1826

Reference ORBIGNY, A.D. D', 1826: Tableau méthodique de la classe des Céphalopodes. — Annales des Sciences Naturelles, Paris, France, Sér. 1, tome 7:277.

Type species *Globigerina bulloides* D'ORBIGNY, 1826.

Diagnosis Test trochospiral.
Wall perforate, surface may be smooth, pitted, cancellated, hispid or spinose.
Chambers spherical to ovate.
Sutures on spiral side radial to curved, on umbilical side radial, depressed.
Aperture interiomarginal, umbilical, with a tendency in some species to extend to a slightly extraumbilical position.

Genus *Globigerinatella* CUSHMAN and STAINFORTH, 1945

Reference CUSHMAN, J.A. and STAINFORTH, R.M., 1945: The Foraminifera of the Cipero Marl Formation of Trinidad, British West Indies. — Cushman Laboratory for Foraminiferal Research, Spec. Publ. 14:68.

Type species *Globigerinatella insueta* CUSHMAN and STAINFORTH, 1945.

Diagnosis Test subglobular, early portion trochospiral with the final chamber usually embracing.
Wall perforate, surface smooth or pitted.
Aperture in the early stage interiomarginal, umbilical, in the later chambers with secondary sutural and areal apertures, surrounded by distinct lips, with small knobby pustule-like bullae covering the areal secondary apertures, or more irregular, spreading, sutural bullae covering the secondary sutural apertures; all bullae may have infralaminal accessory apertures.

Genus *Globigerinatheka* Bronnimann, 1952

Reference Bronnimann, P., 1952: *Globigerinoita* and *Globigerinatheka,* new genera from the Tertiary of Trinidad, B. W. I. — Contributions from the Cushman Foundation for Foraminiferal Research, 3 (1):26.

Type species *Globigerinatheka barri* Bronnimann, 1952.

Diagnosis Test globular, early chambers trochospiral as in *Globigerina,* later with a large enveloping final chamber covering the previous umbilical side as in *Orbulina.*
Wall perforate, surface pitted.
Sutures radial, depressed.
Primary aperture of the early *Globigerina* stage interiomarginal, umbilical, but this is covered in the adult by the final enveloping chamber; the secondary sutural apertures multiple on the spiral side and covered bij small bullae, each of which has one or more small arched infralaminal accessory apertures.

Genus *Globigerinita* Bronnimann, 1951

Reference Bronnimann, P., 1951: *Globigerinita naparimaensis* n.gen., n.sp. from the Miocene of Trinidad, B.W.I. — Contributions from the Cushman Foundation for Foraminiferal Research, 2 (1):18.

Type species *Globigerinita naparimaensis* Bronnimann, 1951.

Diagnosis Test trochospiral.
Wall perforate, surface smooth, pitted or hispid.
Chambers spherical to ovate.
Sutures radial to slightly curved, depressed.
Primary aperture interiomarginal, umbilical, in the final stage completely covered by an irregular bulla expanding along the earlier sutures, with numerous infralaminal accessory apertures along the margins, both at the junction with the sutures of earlier chambers and along the contact with the primary chambers.

Genus *Globigerinoides* CUSHMAN, 1927

Reference
CUSHMAN, J.A., 1927: An outline of a re-classification of the Foraminifera. — Contributions from the Cushman Laboratory for Foraminiferal Research, 3 (1):87

Type species
Globigerina rubra D'ORBIGNY, 1839.

Diagnosis
Test trochospiral.
Wall perforate, surface smooth, hispid or spinose.
Chambers globular to ovate.
Sutures radial to slightly curved, depressed.
Primary aperture interiomarginal, umbilical, with previous apertures remaining open into the umbilicus, smaller secondary sutural apertures on the spiral side, one or more per chamber, often confined to the last few chambers.

Genus *Globigerinoita* BRONNIMANN, 1952

Reference
BRONNIMANN, P., 1952: *Globigerinoita* and *Globigerinatheka,* new genera from the Tertiary of Trinidad, B.W.I. — Contributions from the Cushman Foundation for Foraminiferal Research, 3 (1):26.

Type species
Globigerinoita morugaensis BRONNIMANN, 1952.

Diagnosis
Test trochospiral.
Wall perforate, surface spinose.
Chambers spherical to ovate.
Sutures radial to slightly curved, depressed.
Primary aperture interiomarginal, umbilical, with one or more secondary sutural apertures as in *Globigerinoides* on the spiral side, in the adult stage the primary aperture is covered by an umbilical bulla and the secondary apertures of the spiral side may also be covered by sutural bullae, with commonly two or three accessory infralaminal apertures at the margins of each of the bullae.

Genus *Globoquadrina* FINLAY, 1947

Reference   FINLAY, H. J., 1947: New Zealand Foraminifera: Key Species in Stratigraphy, No. 5 — The N.Z. Journal of Science and Technology, 28 (5):290.

Type species   *Globorotalia dehiscens* CHAPMAN, PARR and COLLINS, 1934.

Diagnosis   Test trochospiral.
Wall perforate, surface pitted to hispid.
Chambers spherical to subangular truncate.
Sutures radial to slightly curved, depressed.
Aperture interiomarginal, umbilical, covered at top by an apertural flap, usually a tooth-like projection.

Genus *Globorotalia* CUSHMAN, 1927

Reference   CUSHMAN, J. A., 1927: An outline of a re-classification of the Foraminifera. — Contributions from the Cushman Laboratory for Foraminiferal Research, 3 (1):91.

Synonymy   *Truncorotalia* CUSHMAN and BERMUDEZ, 1949: Contributions from the Cushman Laboratory for Foraminiferal Research, 25 (2):35.
Turborotalia CUSHMAN and BERMUDEZ, 1949: Contributions from the Cushman Laboratory for Foraminiferal Research, 25 (2):49.
Acarinina SUBBOTINA, 1953: Trudy Vses. Neft. Naukno — Issledov. Geol. — Razved. Inst., 76:219.
Planorotalites MOROZOVA, 1957: Dokl. Acad. Sci. U.R.S.S., 114:1112.

Type species   *Pulvinulina menardii* D'ORBIGNY var. *tumida* BRADY, 1877.

Diagnosis   Test trochospiral, biconvex to umbilicoconvex, umbilicate, periphery with or without a single keel.
Wall perforate, surface smooth to finely spinose or slightly nodose.
Chambers ovate to angular rhomboid or angular conical.
Sutures on the spiral side depressed to elevated, curved or radial, on the umbilical side usually depressed, radial.
Aperture interiomarginal, an extraumbilical-umbilical arch bordered by a lip, varying from a narrow rim to a spatulate flap.

Genus *Hantkenina* CUSHMAN, 1924

Reference CUSHMAN, J. A., 1924: A new Genus of Eocene Foraminifera. — Proceedings United States National Museum, 66 (30):1.

Synonymy *Sporohantkenina* BERMUDEZ, 1937: Mem. Soc. Cubana Hist. Nat., 11:151.
Aragonella THALMANN, 1942: Amer. Journ. Sci., 240:811.
Applinella THALMANN, 1942: Amer. Journ. Sci., 240:812.
Hantkeninella BRONNIMANN, 1950: Journ. Paleontol., 24 (4):399.

Type species *Hantkenina alabamensis* CUSHMAN, 1924.

Diagnosis Test planispiral, biumbilicate, almost involute, biconvex.
Wall perforate, surface finely pitted or smooth, may be finely hispid in the area just beneath the aperture on the previous whorl.
Chambers rounded, ovate or radially elongate, compressed to inflated, with a single distinct peripheral tubulospine at the forward margin of each chamber.
Sutures radial to slightly curved, depressed.
Aperture interiomarginal, equatorial, triradiate, two of the rays forming a slit across the base of the final chamber face, the third ray originating from the centre of this slit and extending up the face towards the peripheral spine, the rays bordered by an apertural flange.

Genus *Hastigerina* THOMSON, 1876

Reference THOMSON, W., 1876: Preliminary reports to Professor Wyville Thomson, F.R.S., director of the Civilian Scientific Staff, on work done on board the "Challenger". — Proceedings Royal Society London, 24:534.

Synonymy *Globigerinella* CUSHMAN, 1927: Contributions from the Cushman Laboratory for Foraminiferal Research, 3:87.

Type species *Hastigerina murrayi* THOMSON, 1876.

Diagnosis Test planispiral, early stage may be slightly trochospiral, biumbilicate, involute to loosely coiled; lobulate in outline, no keel present.
Wall finely to coarsely perforate, surface smooth, hispid or spinose.
Chambers spherical to ovate.
Sutures radial, depressed.
Aperture interiomarginal, a broad equatorial arch.

Genus *Hastigerinella* CUSHMAN, 1927

Reference CUSHMAN, J. A., 1927: An outline of a re-classification of the Foraminifera. — Contributions from the Cushman Laboratory for Foraminiferal Research, 3 (1):87.

Type species *Hastigerina digitata* RHUMBLER, 1911.

Diagnosis Test trochospiral.
Chambers globular in early stage, later chambers radially elongate, clavate or cylindrical.
Wall perforate, surface smooth except at the outer ends of the chambers where spines may be concentrated; the latter, however, are usually broken in fossil shells.
Sutures radial, depressed.
Aperture interiomarginal, extraumbilical-umbilical, a broad arch, extending to the periphery or even becoming spiroumbilical.

Genus *Orbulina* D'ORBIGNY, 1839

Reference ORBIGNY, A. D. D', 1839: Foraminifères *in* de la Sagra, Histoire physique, politique et naturelle de l'Ile de Cuba. — p.2.

Type species *Orbulina universa* D'ORBIGNY, 1839.

Diagnosis Test generally spherical, rarely two- or three-chambered, early stage trochospiral, in the adult the final chamber or rarely the penultimate chamber completely or partially enveloping the globigerine coil.
Wall coarsely perforate, surface smooth to pitted.
Primary aperture interiomarginal, umbilical in the early globigerine stage, in the adult areal with numerous small openings, which may be scattered over the test; small sutural secondary openings are commonly found around the early globigerine chambers of specimens where these are visible at the surface.

Genus *Orbulinoides* Blow and Saito, 1968

Reference Blow, W. H. and Saito, T., 1968: The morphology and taxonomy of *Globigerina mexicana* Cushman, 1925. – Micropaleontology, 14 (3): 357.

Type species *Porticulasphaera beckmanni* Saito, 1962.

Diagnosis Test subglobular to globular, early portion trochospiral, final chamber much inflated to almost spherical, embracing the greater proportion of the earlier test.
Wall comparatively thick, coarsely perforate, surface pitted.
Sutures radial to slightly curved, depressed.
Primary aperture in the early portion interiomarginal, umbilical, with secondary sutural openings on the spiral side, as in *Globigerinoides*, the umbilical aperture covered by the final embracing chamber of the adult, which has numerous small secondary apertures along the basal suture; these, together with the secondary sutural apertures on the spiral side, remain uncovered.

Genus *Praeorbulina* Olsson, 1964

Reference Olsson, R. K., 1964: *Praeorbulina* Olsson, a new foraminiferal genus. – Journal of Paleontology, 38 (4): 770.

Type species *Globigerinoides glomerosa glomerosa* Blow, 1956.

Diagnosis Test globular to spherical or almost bispherical, early portion trochospiral, only final chamber embracing the greater proportion of the earlier test or penultimate one already embracing part of the earlier test.
Wall distinctly perforate, surface pitted.
Sutures slightly curved to radial, slightly depressed.
Primary aperture in the early portion interiomarginal, umbilical, with secondary sutural openings on the spiral side, as in *Globigerinoides*. The final chamber has three low arched or several small, slitlike openings along the basal suture.

Genus *Pulleniatina* CUSHMAN, 1927

Reference CUSHMAN, J. A., 1927: An outline of a re-classification of the Foraminifera. — Contributions from the Cushman Laboratory for Foraminiferal Research, 3 (1):90.

Type species *Pullenia obliquiloculata* PARKER and JONES, 1865.

Diagnosis Test globose, trochospiral to streptospiral, early portion as in *Globigerina,* with open umbilicus, later chambers completely enveloping the entire umbilical side of the previous trochospiral coil, including the previous open umbilicus; it thus may even appear involutely-coiled.
Wall perforate, later part comparatively thickened, surface in the adult smooth, although the portion of the earlier whorl just below the aperture may show a hispid surface.
Sutures slightly depressed, radial to slightly curved in the *Globigerina* stage.
Aperture interiomarginal, in the young a broad umbilical arch, as in *Globigerina,* in the adult a broad low extraumbilical arch at the base of the final enveloping chamber, bordered above by a thickened lip, but not directly opening into the earlier umbilicus because of the streptospiral growth pattern.

Genus *Sphaeroidinella* CUSHMAN, 1927

Reference CUSHMAN, J. A., 1927: An outline of a re-classification of the Foraminifera. — Contributions from the Cushman Laboratory for Foraminiferal Research, 3 (1):90.

Type species *Sphaeroidina dehiscens* PARKER and JONES, 1865.

Diagnosis Test elongate ovate, early portion trochospiral, the two or three much-embracing chambers of the final whorl enveloping the early whorl, each with marginal flanges extending out towards the opposing chambers and partially obscuring the arched apertures.
Primary wall coarsely perforate, covered by secondary layers of shell material greatly reducing the external openings of the pores of the primary wall or completely sealing them; surface smooth and glassy in appearance.
Sutures radial or curved, depressed.
Primary aperture in the early stage interiomarginal, umbilical, in the adult covered by the embracing final chamber; one or more sutural secondary apertures on opposite sides of the final chamber may be present. All apertures are furnished with thickened lips, which may be smooth or crenulate and may become flange-like.

Genus *Truncorotaloides* BRONNIMANN and BERMUDEZ, 1953

Reference BRONNIMANN, P. and BERMUDEZ, P. J., 1953: *Truncorotaloides,* a new Foraminiferal genus from the Eocene of Trinidad, B.W.I. — Journal of Paleontology, 27 (6):817.

Type species *Truncorotaloides rohri* BRONNIMANN and BERMUDEZ, 1953.

Diagnosis Test trochospiral, biconvex to umbilicoconvex, umbilicate, with or without a single keel. Wall perforate, surface prominently hispid to spinose.
Chambers ovate to angular-rhomboid.
Sutures on the spiral side radial to oblique, depressed, on the umbilical side radial, depressed.
Primary aperture interiomarginal, extraumbilical-umbilical, with secondary sutural apertures on the spiral side at the inner margin of the later chambers.

DESCRIPTIONS OF THE CENOZOIC ZONES

The present zonation of the Cenozoic is largely based on the work of H. M. BOLLI (1957, 1966). However, earlier investigations carried out by industrial palaeontologists in the Caribbean, like T. F. GRIMSDALE (1951), should also be mentioned. Since 1955 the author has had the opportunity to study the planktonic faunas from other areas such as Italy, Turkey, West Pakistan, northwest Borneo, Indonesia, Philippines, etc. and has compared his own observations with the information available from the Caribbean area. W. H. BLOW, in several publications since 1956, greatly contributed to the refinement of our knowledge and recently presented the documentation of his detailed zonation (1969), which covers the interval Late Middle Eocene–Recent.

The information available on the time-stratigraphic distribution of the planktonic Foraminifera in the Early Cenozoic is still considerably more extensive than that on the Late Cenozoic. This is due to the fact that most of the younger marine sediments still form part of the present sea bottom and hence are not easily accessible. Moreover, the diversity of characteristic short-ranging planktonic species is almost exuberant during the Early Cenozoic interval.

It is outside the scope of this manual to deal with all time-stratigraphical problems which are related to this zonation. The correlation of the "warm-water" planktonic foraminiferal zones with the time-honoured but unpractical European time scale is impaired by the unfavourable facies of most of the type localities of the stages. It becomes even more problematic in the Late Cenozoic due to contraction of the tropical and sub-tropical belt. An example of the confusion arising from the foregoing considerations is the lack of unanimity amongst authors as to what exactly comprises the Oligocene in Central America, or even if it is present at all (EAMES et al., 1962).

Within the tropical belt itself other environmental aspects may be the cause of bio-stratigraphic misinterpretations. The planktonic zonation devised by BLOW and BANNER (1962), and further elaborated in later papers (BLOW and BANNER, 1965; BLOW, 1969), may be cited as an example. A number of zones were introduced (1962), based on work in the Lindi area of Tanzania. Two of these, later designated P18–19 (1965), were stated to be of Oligocene age and to represent the missing interval in Central America. They are here considered unacceptable because they are based on what appear to be atypical planktonic assemblages (op. cit. 1962, pp.67, 69). In my view these zones and zone P17 are the approximate equivalent of the Upper P16 and N1 interval of these authors.

On charts 2 and 3 the positions of the epochs relative to the planktonic foraminiferal zonation are indicated, although it is realized that they will be liable to minor shifts. The definitions of the zones are given below:

17 **Globigerina daubjergensis** zone (Total-range zone)

This zone corresponds to the *Globorotalia trinidadensis* zone as established in Trinidad (BOLLI, 1957). It contains the earliest Cenozoic planktonic Foraminifera: the zonal marker, which is restricted to it, and *G. triloculinoides*, *Globorotalia trinidadensis*, *Globorotalia compressa* and *Globorotalia pseudobulloides*.

18 **Globorotalia uncinata** zone (Partial-range zone)

The zone is defined as that part of the range of *G. uncinata* prior to the first appearance of *G. ehrenbergi*, *G. abundocamerata* and *G. pusilla*. Transitional forms between *G. uncinata* and *G. angulata* begin to appear in the upper part of this zone.

19 Globorotalia angulata zone (Concurrent-range zone)
The concurrence of the stratigraphic ranges of *G. angulata*, *G. trinidadensis*, *G. compressa*, *Globigerina triloculinoides* and *G. pusilla*, *G. ehrenbergi* and *G. abundocamerata* characterizes this zone. In this zone the first keeled *Globorotalia* species appear.

20 Globorotalia pseudomenardii zone (Total-range zone)
The zone is characterized by the zonal marker. Also restricted to the zone are *G. laevigata* and *G. mckannai*. Very characteristic and easily recognizable is *G. velascoensis*, which species continues into the overlying zone.
G. ehrenbergi and *G. abundocamerata* become extinct within the zone.

21 Globorotalia velascoensis zone (Concurrent-range zone)
The zone is defined as the range overlap of *G. velascoensis* and *Globigerina soldadoensis*. The absence of *G. pseudomenardii*, *G. laevigata* and *G. mckannai* is an important negative feature.

22 Globorotalia rex zone (Partial-range zone)
This zone is defined as that part of the range of *G. rex* below the first appearance of *G. formosa* and *G. aragonensis*. *G. wilcoxensis* is restricted to the zone.

23 Globorotalia formosa/aragonensis zone (Partial-range zone)
This zone is a combination of the *G. formosa* and *G. aragonensis* zones as established in Trinidad (BOLLI, 1957).
It was found that in certain areas the two zones may be difficult to distinguish from each other, as has for instance been found to be the case in northern Italy, where they were combined by BOLLI and CITA (1960). *G. formosa* and *G. aragonensis* both appear in this zone, while *G. formosa* is restricted to it. *G. aragonensis* continues into younger zones.

24 Globorotalia bullbrooki zone (Partial-range zone)
The zone is defined as that part of the ranges of *G. bullbrooki*, *G. spinulosa*, *G. renzi*, *G. bolivariana*, *Hantkenina aragonensis*, *Globigerina senni*, *Globigerina boweri*, *Clavigerinella akersi* and *C. jarvisi*, which occur before the first appearance of *Hantkenina dumblei*, *Globigerapsis index* and *Globigerina yeguaensis*. *G. aspensis* is an important species in this zone.

25 Globigerapsis kugleri zone (Concurrent-range zone)
The concurrence of the stratigraphic ranges of *Globorotalia broedermanni*, *Globorotalia aragonensis*, *Globigerina boweri*, *Hantkenina mexicana* and *Hantkenina dumblei*, *G. index* and *Globigerina yeguaensis* characterizes the zone. For the sake of continuity the zone is named after the very characteristic *G. kugleri*.

26 Globorotalia lehneri zone (Partial-range zone)
The part of the stratigraphic range of the zonal marker, which extends above the top occurrence of *G. aragonensis*, *G. broedermanni*, *Globigerina boweri*, *Clavigerinella akersi* and *Hantkenina mexicana* and below the base occurrence of *Orbulinoides beckmanni* constitutes the zone.
This zone is intermediate between the readily recognizable *Globigerapsis kugleri* and *Orbulinoides beckmanni* zones. Notable is the first appearance of *Globigerinatheka barri*.

27 Orbulinoides beckmanni zone (Total-range zone)
The zonal marker has a very characteristic morphology which makes this zone

easily identifiable. Moreover, *Globorotalia spinulosa*, *Truncorotaloides topilensis* and *Globigerapsis kugleri* become extinct at the top of the zone.

28   Truncorotaloides rohri zone (Partial-range zone)
This zone is defined as that part of the stratigraphic range of *T. rohri* prior to the first appearance of *Globigerapsis mexicana* and *Globorotalia cerroazulensis* and after the last appearance of *G. spinulosa*, *T. topilensis*, *Globigerapsis kugleri* and *Orbulinoides beckmanni*. *Globigerina ouachitaensis* makes its first appearance at the base of this zone, while several species become extinct at its top.

29   Globigerapsis mexicana zone (Total-range zone)
The zone is characterized by the restricted occurrence of the zonal marker. *Hantkenina brevispina* and *Globorotalia cerroazulensis* have their first appearance at the base of the zone.
Transitional forms of *Globorotalia centralis* to *G. cerroazulensis* occur.

30   Globorotalia cerroazulensis zone (Concurrent-range zone)
The zone, of uppermost Eocene age, is characterized as the stratigraphic range overlap of *G. cerroazulensis*, *G. centralis* and *Globigerina ampliapertura*. The last *Hantkenina* species disappear near or at the top of the zone; restricted to the zone is *Cribrohantkenina bermudezi*.

31   Globigerina ampliapertura zone (Partial-range zone)
That part of the range of the zonal marker, which extends above the top occurrence of *Globorotalia centralis*, *Globorotalia cerroazulensis* and *Cribrohantkenina bermudezi* and below the base occurrence of *Globorotalia opima* and *Globigerina angulisuturalis* constitutes the zone. *Cassigerinella chipolensis* makes its first appearance at the base of this zone. *G. ciperoensis*, one of the most valuable guide fossils at this level, makes its first appearance at or near the base of the zone.
An important feature of Oligocene planktonic associations in general is the absence of any keeled or spinose species, which makes it easy to distinguish them at first glance from Eocene associations.

32   Globigerina angulisuturalis zone (Partial-range zone)
This zone is defined as that part of the range of *G. angulisuturalis* prior to the first appearance of *Globorotalia kugleri*. The planktonic association is characterized by the co-occurrence of *G. angulisuturalis*, *G. ciperoensis*, *G. sellii* and *G. venezuelana*. The lower part of the zone may be distinguished by the occurrence of *Globorotalia opima*.

33   Globorotalia kugleri zone (Partial-range zone)
The zone is defined as that part of the range of *G. kugleri*, which occurs before the first appearance of *Globigerinoides trilobus* and *Globigerinoides immaturus*. Valuable species in this zone are *Globigerinoides primordius* and *Globigerina binaiensis*. *Globigerina angulisuturalis* is useful to indicate the lower part of the zone.

34   Globigerinoides trilobus zone (Concurrent-range zone)
This zone is defined as the stratigraphic range overlap of *G. trilobus*, *G. immaturus* and *Catapsydrax dissimilis*. The zone has been established by combining the *Catapsydrax stainforthi*, *C. dissimilis* and uppermost part of the *Globorotalia kugleri* zones of BOLLI, 1957. *Globigerinatella insueta*, where present with *C. dissimilis*, is an important marker for the upper part of the *G. trilobus* zone. Similarly the last occurrence of *Globigerina tripartita* appears to be in the lower part of this zone.

35   Globigerinatella insueta zone (Partial-range zone)
That part of the stratigraphic range of *G. insueta*, which extends above the top occurrence of *Catapsydrax dissimilis* and below the base occurrence of *Orbulina*

universa, *O. suturalis* and *O. bilobata*, constitutes the zone. *Globigerinoides diminutus* makes its first appearance at the base of this zone. *Globigerinoides sicanus* without *Orbulina s.l.* is a useful marker for the upper part of the zone.

36 Globorotalia peripheroronda zone (Concurrent-range zone)
This zone is defined as the stratigraphic range overlap of *G. peripheroronda*, *G archeomenardii*, *Orbulina bilobata*, *O. suturalis* and *O. universa*. Useful species in the zonal association are *Praeorbulina transitoria*, *P. glomerosa s.l.* and, if present, *Hastigerinella bermudezi*. The base of the zone is the well known *Orbulina* datum.

37 Globorotalia peripheroacuta zone (Partial-range zone)
That part of the stratigraphic range of *G. peripheroacuta*, which extends above the top occurrence of *G. peripheroronda* and *G. archeomenardii* defines the zone.

38 Globorotalia lobata zone (Partial-range zone)
The lower boundary of the zone is characterized by the top occurrence of *G. peripheroacuta*; the upper boundary of the zone coincides with the first occurrence of *G. fohsi*.

39 Globorotalia fohsi zone (Total-range zone)
The zone is characterized by the restricted stratigraphic range of the zonal marker, by the presence of *G. praemenardii* and *G. mayeri*, while the absence of *Sphaeroidinella subdehiscens* may be an important negative feature.

40 Globigerinoides subquadratus zone (Partial-range zone)
That part of the stratigraphic range of the zonal marker, which extends above the top occurrence of *Globorotalia fohsi* and *Globorotalia praemenardii*. *Sphaeroidinella subdehiscens* has its first appearance at the base of this zone; *Globorotalia mayeri* becomes extinct in the upper part of the zone.

41 Globorotalia siakensis zone (Partial-range zone)
This zone is defined as that part of the stratigraphic range of *G. siakensis*, which occurs after the disappearance of *Globigerinoides subquadratus*. *Globigerina nepenthes* probably starts already in the lower part of the zone, where *Hastigerina aequilateralis* still is absent.

42 Globorotalia menardii zone (Partial-range zone)
The zone is defined as that part of the stratigraphic range of *G. menardii*, which extends above the top occurrence of *G. siakensis* and below the base occurrence of *G. acostaensis*. The zonal assemblage still includes *G. lenguaensis*, the occurrence of *G. pseudomiocenica* is questionable.

43 Globorotalia acostaensis zone (Partial-range zone)
That part of the stratigraphic range of the zonal marker, which occurs prior to the first appearance of *G. plesiotumida* and *G. dutertrei* defined the zone. *Globigerinoides extremus*, a very characteristic species, is indicative for the upper part of the zone.

44 Globorotalia dutertrei zone (Partial-range zone)
The zone is defined as that part of the stratigraphic range of *G. dutertrei*, which extends prior to the base occurrences of *G. margaritae* and *G. tumida*. *Pulleniatina primalis*, *G. miocenica* and *G. multicamerata* probably have their first appearances in the uppermost part of this zone.

45 Globorotalia margaritae zone (Total-range zone)
The zone is based on the restricted stratigraphic range of *G. margaritae*. *G. tumida* has its first appearance at the base of the zone. *G. pseudomiocenica* probably becomes extinct in the middle part of the zone, the first occurrences of *Globigerina riveroae*,

Globorotalia crassaformis and *Sphaeroidinella dehiscens* indicate the upper part of the zone.

46   Globoquadrina altispira zone (Partial-range zone)
That part of the stratigraphic range of the zonal marker, which extends above the top occurrence of *Globorotalia margaritae* constitutes this zone. *Globigerina nepenthes* and *Globigerinoides extremus* become extinct in the upper part of the zone, *Globigerina riveroae* does so at the upper boundary of this zone, while *Globigerinoides fistulosus*, *Pulleniatina obliquiloculata* and *Globorotalia tosaensis* have their first appearances in the upper or uppermost part of the zone.

47   Globorotalia tosaensis zone (Partial-range zone)
That part of the stratigraphic range of the zonal marker, which extends above the top occurrence of *Globoquadrina altispira*, *Sphaeroidinella subdehiscens* and *Globigerina riveroae* and below the base occurrence of *G. truncatulinoides* constitutes the zone. *G. multicamerata* and *Pulleniatina praecursor* in the absence of the above three species are indicative for the lower part of the zone.

48   Globorotalia truncatulinoides zone (Total-range zone)
This zone is exclusively defined by the restricted stratigraphic range of the zonal guide fossil.

Paleocene-Eocene

DESCRIPTIONS AND ILLUSTRATIONS OF PALEOCENE-EOCENE SPECIES
(arranged in alphabetical order)

Clavigerinella jarvisi (CUSHMAN)

Reference　*Hastigerinella jarvisi* CUSHMAN, 1930: Fossil species of *Hastigerinella*. — Contributions from the Cushman Laboratory for Foraminiferal Research, 6 (1) :18, pl.3, figs. 8-11.

Type locality　17 miles out on Cunapo Southern road, Trinidad.

Diagnosis　Test planispiral, biumbilicate, involute, lobulate in outline.
Wall distinctly perforate, surface finely pitted.
Early chambers spherical, later chambers radially elongate, but not inflated at the outer ends; the four chambers of the last whorl increase rapidly in size.
Sutures radial, depressed.
Aperture interiomarginal, equatorial, an elongate slit extending up the apertural face for about half the length of the final chamber, bordered laterally by wide flanges which are flared at the base and join at the top.

Strat. distr.　Base of *Globorotalia bullbrooki* zone to top of *Truncorotaloides rohri* zone.
Questionable occurrence in the *Globigerapsis mexicana* zone.

Clavigerinella akersi BOLLI, LOEBLICH and TAPPAN

Reference　*Clavigerinella akersi* BOLLI, LOEBLICH and TAPPAN, 1957: Planktonic Foraminiferal Families Hantkeninidae, Orbulinidae, Globorotaliidae and Globotruncanidae. — United States National Museum Bulletin, 215:30, pl.3, fig. 5a-b.

Type locality　1100 feet south of the 12.5 milepost of the Brasso-Tamana Road, Eocene Navet Formation, Trinidad.

Diagnosis　Test planispiral, biumbilicate, involute, lobulate in outline.
Wall distinctly perforate, surface finely pitted.
Early chambers spherical, later chambers radially elongate and characteristically greatly inflated at the outer ends; the four chambers of the last whorl increase rapidly in size.
Sutures radial, depressed.
Aperture as in *Clavigerinella jarvisi*.

Strat. distr.　Base of *Globorotalia bullbrooki* zone to top of *Globigerapsis kugleri* zone

Remarks　*Clavigerinella akersi* BOLLI, LOEBLICH and TAPPAN is distinguished from *Clavigerinella jarvisi* (CUSHMAN) by having the later, elongate chambers distinctly inflated at the outer ends.
Clavigerinella akersi is closely related to *C. colombiana* (PETTERS, 1954); *C. jarvisi* is most probably closely related to *C. eocanica* (NUTTALL, 1928) and *C. eocanica* var. *aragonensis* (NUTTALL, 1930).

× 65

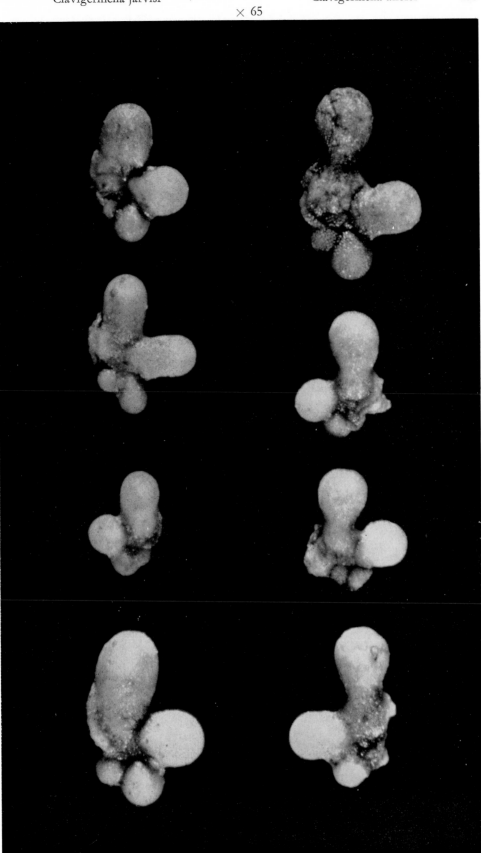

Cribrohantkenina bermudezi THALMANN

Reference   *Hantkenina (Cribrohantkenina) bermudezi* THALMANN, 1942: Foraminiferal genus *Hantkenina* and its subgenera. — American Journal of Science, 240:812, pl.1, fig. 5.

Type locality   Palmer Sta. 1640, just north of Grua 9, Ramal Juan Criollo of Central Jatibonico, Camaguay Province, Cuba.

Diagnosis   Test planispiral, biumbilicate, almost involute; equatorial periphery slightly lobulate to almost circular, axial periphery rounded.
Wall finely perforate, surface smooth.
Chambers inflated, subglobular, with a prominent peripheral tubulospine at the forward margin of each chamber; the four to five succeeding chambers of the last whorl are attached near the base of the spines and may partially or completely envelop the spine of the preceding chamber.
Sutures almost radial, depressed.
Primary aperture an interiomarginal, equatorial arch, projecting into the umbilical regions, bordered above by a distinct lip; secondary apertures are areal supplementary apertures (two or more) arranged in one or more rows above the primary aperture, bordered by distinct, protruding lips.

Strat. distr.   Ranging throughout the *Globorotalia cerroazulensis* zone.

Remarks   The following species are closely related or synonyms:
Hantkenina inflata HOWE, 1928
Hantkenina mccordi HOWE and WALLACE, 1932
Hantkenina danvillensis HOWE and WALLACE, 1934
Locality of figured specimen is the above-mentioned type locality.

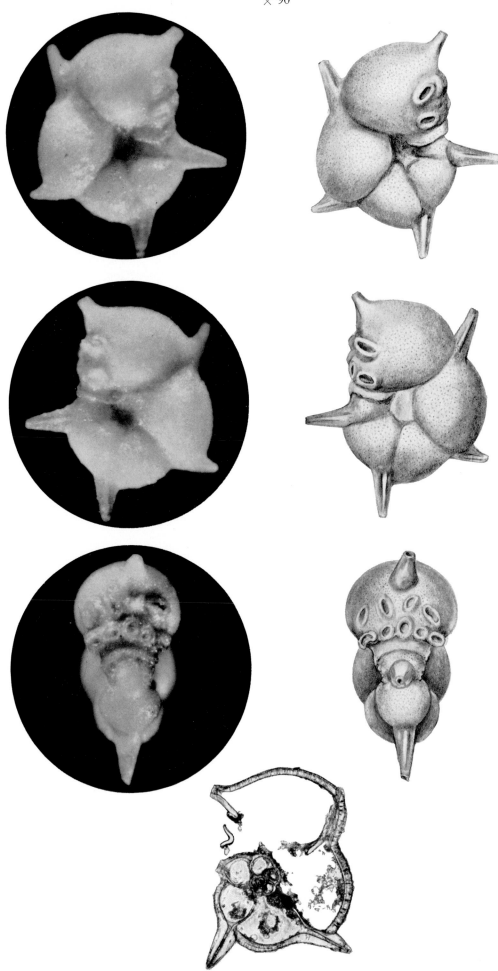

Globigerapsis index (FINLAY)

Reference   *Globigerinoides index* FINLAY, 1939: New Zealand Foraminifera: Key Species in Stratigraphy — No. 2 — Transactions Royal Society of New Zealand, 69:125, pl.14, figs. 85-88.

Type locality   Locality 5179A, beach 1 mile north of Kakaho Creek, Hampden section, South Island, New Zealand.

Diagnosis   Test subglobular, early portion trochospiral.
Wall coarsely perforate, surface of early chambers pitted and sometimes slightly spinose or nodose.
Chambers subglobular, rapidly increasing in size, final chamber covering the umbilicus. Sutures deeply cleft, slightly curved to radial.
Aperture in the early stage interiomarginal, umbilical, but covered in the adult by the final embracing chamber, which has three sutural secondary apertures at the basal margin, each opposite a suture; at least the final aperture is very distinctly high-arched.

Strat. distr.   Base of *Globigerapsis kugleri* zone into lowest part of *Globorotalia cerroazulensis* zone.

Remarks   This species differs from *Globigerapsis kugleri* BOLLI, LOEBLICH and TAPPAN in having a less or smaller embracing final chamber, in having fewer chambers per whorl and higher-arched apertures. Locality of figured specimen is the above-mentioned type locality.

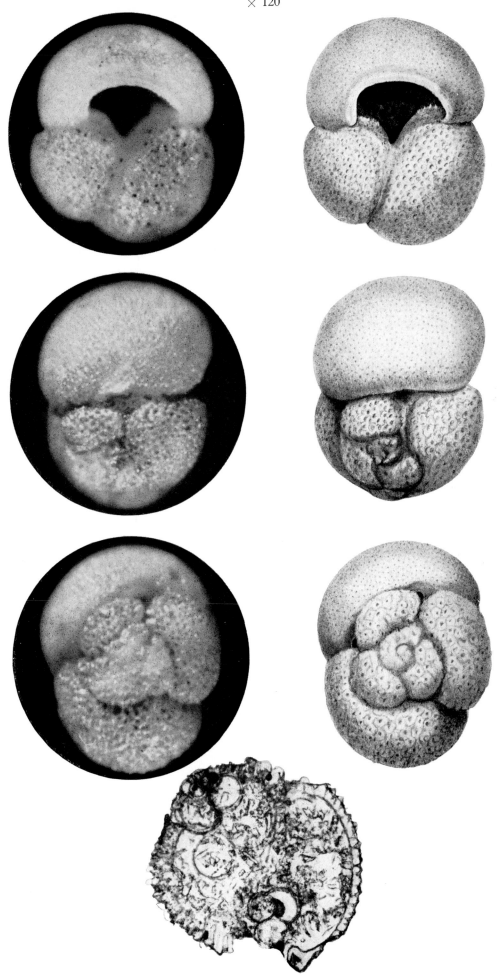

Globigerapsis kugleri BOLLI, LOEBLICH and TAPPAN

Reference   *Globigerapsis kugleri* BOLLI, LOEBLICH and TAPPAN, 1957: Planktonic Foraminiferal Families Hantkeninidae, Orbulinidae, Globorotaliidae and Globotruncanidae. — United States National Museum Bulletin, 215:34, pl.6, fig. 6a-c.

Type locality   Holotype from the Eocene Navet Formation, Penitence Hill marl, from a block in the Oligocene Nariva Formation, Pointe-à-Pierre, Trinidad.

Diagnosis   Test subglobular, early portion trochospiral.
Wall coarsely perforate, surface slightly spinose or nodose.
Chambers of early portion globular, rapidly increasing in size, last chambers compressed, final chamber considerably larger, somewhat embracing, covering the umbilical region of the early coil.
Sutures depressed, commonly almost incised, slightly curved to radial.
Aperture in the early stage interiomarginal, umbilical, but covered in the adult by the final embracing chamber, which has several (two to four) arched sutural secondary apertures at the basal margin opposite a suture.

Strat. distr.   Upper part of *Globigerapsis kugleri* zone to top of *Orbulinoides beckmanni* zone. Questionable occurrence in lower part of *Globigerapsis kugleri* zone.

Remarks   This species differs from *Globigerapsis mexicana* (CUSHMAN) in having more inflated chambers, more deeply incised sutures, a somewhat less embracing final chamber and lower and less arched secondary apertural openings.
Locality of figured specimen is DB 129, Trinidad.

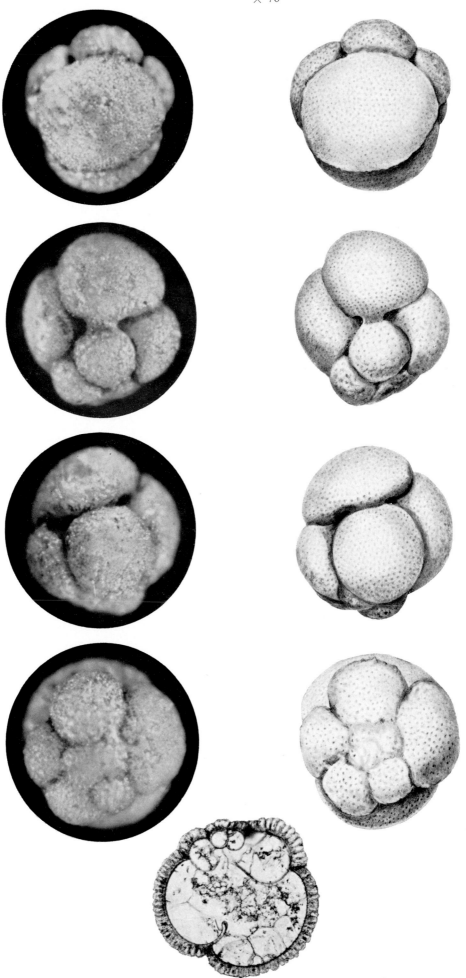

Globigerapsis mexicana (Cushman)

Reference *Globigerina mexicana* Cushman, 1925: New foraminifera from the Upper Eocene of Mexico. – Contributions from the Cushman Laboratory for Foraminiferal Research, 1 (3): 6, pl. 1, figs. 8a-b.

Type locality Palacho Hacienda, S. of Panuco-Tampico R.R., State of Vera Cruz, Mexico.

Diagnosis Test subglobular, early portion trochospiral.
Wall coarsely perforate, surface smooth.
Chambers subglobular, rapidly increasing in size, final chamber covering the umbilicus, embracing a considerable part of the test.
Sutures slightly depressed, slightly curved to radial.
Aperture in the early stage interiomarginal, umbilical, but covered in the adult by the final embracing chamber, which has three distinctly rounded sutural secondary apertures at the basal margin.

Strat. distr. Ranging throughout the *Globigerapsis mexicana* zone.

Remarks *Globigerinoides semiinvoluta* Keyzer, 1945, appears to be a synonym of *Globigerapsis mexicana*.
See remarks *Globigerapsis kugleri* Bolli, Loeblich and Tappan.
Locality of figured specimen is DB 229, Trinidad.

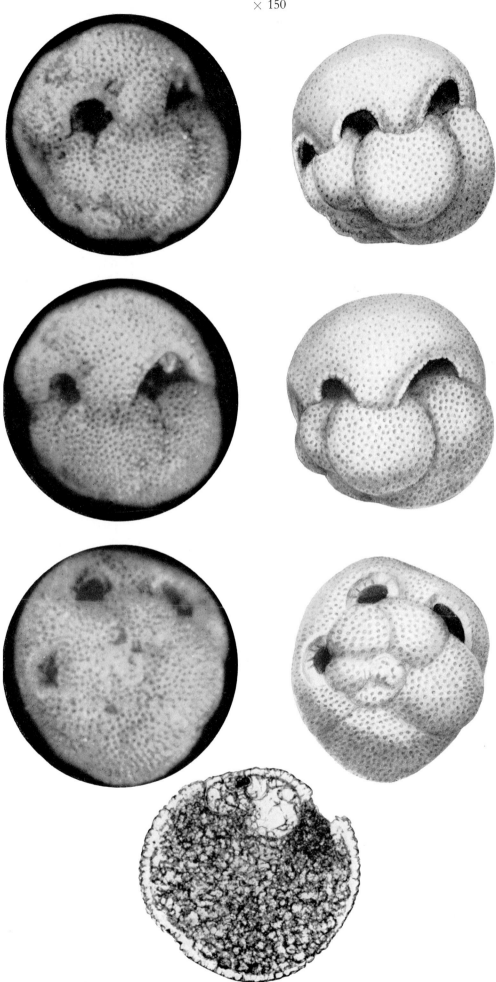

Globigerina ampliapertura BOLLI

Reference   *Globigerina ampliapertura* BOLLI, 1957: Planktonic Foraminifera from the Oligocene-Miocene Cipero and Lengua Formations of Trinidad, B.W.I. — United States National Museum Bulletin, 215:108, pl.22, figs. 4a-7b.

Type locality   Holotype (fig. 6a-c) from surface sample of the northernmost position of the Cipero type section, Trinidad.

Diagnosis   Test low trochospiral, spiral side almost flat to slightly convex, umbilical side convex; equatorial periphery moderately lobulate, axial periphery broadly rounded.
Wall perforate, surface pitted in early chambers to finely pitted in last chamber.
Chambers subspherical, somewhat compressed laterally, arranged in about three whorls; the chambers of the last whorl, usually four in number, increase fairly rapidly in size.
Sutures on spiral side slightly curved to radial, depressed; on umbilical side radial, depressed.
Umbilicus fairly narrow.
Aperture a high, distinct arch, bordered by a rim, interiomarginal, umbilical.

Strat. distr.   Base of *Globorotalia cerroazulensis* zone into lowermost part of *Globigerina angulisuturalis* zone.

Remarks   *Globigerina ampliapertura* BOLLI and *Globorotalia increbescens* (BANDY) are closely related; they differ mainly in the position of the aperture. Locality of figured specimen is Bg 49, Trinidad.

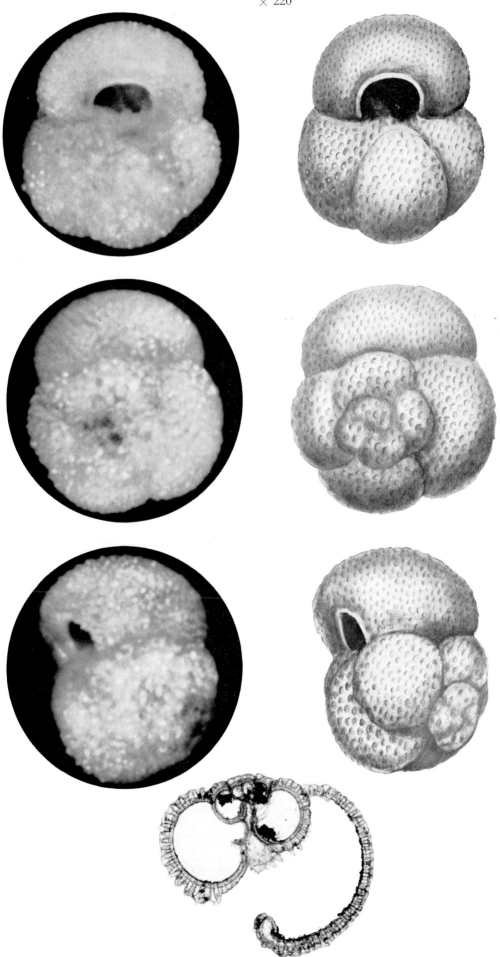

Globigerina boweri BOLLI

Reference	*Globigerina boweri* BOLLI, 1957: Planktonic Foraminifera from the Eocene Navet and San Fernando Formations of Trinidad, B.W.I. — United States National Museum Bulletin, 215:163, pl.36, figs. 1a-2b.
Type locality	Holotype from an outcrop on left side of right branch of Nariva River, about 450 feet from its junction, Central Range, Trinidad. Navet Formation.
Diagnosis	Test low trochospiral; equatorial periphery lobulate, axial periphery broadly rounded. Wall coarsely perforate, surface reticulate. Chambers spherical, early ones somewhat compressed, arranged in about $2\frac{1}{2}$ whorls; the 3-$3\frac{1}{2}$ chambers of the last whorl increase very rapidly in size. Sutures on spiral side slightly curved, depressed; on umbilical side radial, depressed. Umbilicus narrow. Aperture a large, high, interiomarginal, umbilical arch, bordered by a short lip or rim.
Strat. distr.	Base of *Globorotalia bullbrooki* zone to top of *Globigerapsis kugleri* zone.
Remarks	The main characteristic of this species is its high-arched aperture. Locality of figured specimen is Tschopp 72, Cuba.

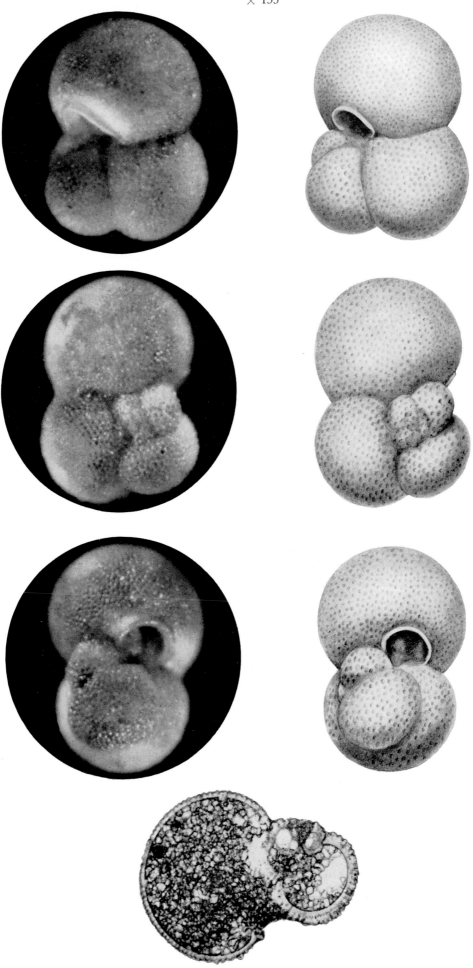

Globigerina collactea (FINLAY)

Reference   *Globorotalia collactea* FINLAY, 1939: New Zealand Foraminifera: Key Species in Stratigraphy — No. 3 — Trans. Royal Society of New Zealand, 69 (3):327, pl.29, figs. 164-165.

Type locality   Locality 5540, Hampden beach section, North Otago, South Island, New Zealand.

Diagnosis   Test low trochospiral, spiral side almost flat to slightly convex, umbilical side convex; equatorial periphery lobulate, may be somewhat elongate, axial periphery rounded.
Wall coarsely perforate, surface covered with minute papillae.
Chambers inflated, subglobular, arranged in about $2\frac{1}{2}$ whorls; the five, or sometimes four, chambers of the last whorl increase fairly rapidly in size.
Sutures on spiral side curved, depressed; on umbilical side radial, depressed.
Umbilicus fairly narrow.
Aperture a low, interiomarginal, umbilical arch, bordered by a rim or faint lip. Often a slight shifting of the aperture towards an extraumbilical-umbilical position is noted.

Strat. distr.   Base of *Globorotalia rex* zone to top of *Globorotalia formosa/aragonensis* zone.

Remarks   Some doubt exists as to the generic position of this species; FINLAY (1939) originally described it as a *Globorotalia*. Because of the umbilical position of the aperture, BRONNIMANN (1952) removed it to *Globigerina*. The apertures of the specimens examined are usually umbilical, though a slight shifting of the aperture of the ultimate chamber towards an extraumbilical-umbilical position is often noted.
Locality of figured specimen is HK 1831, Trinidad.

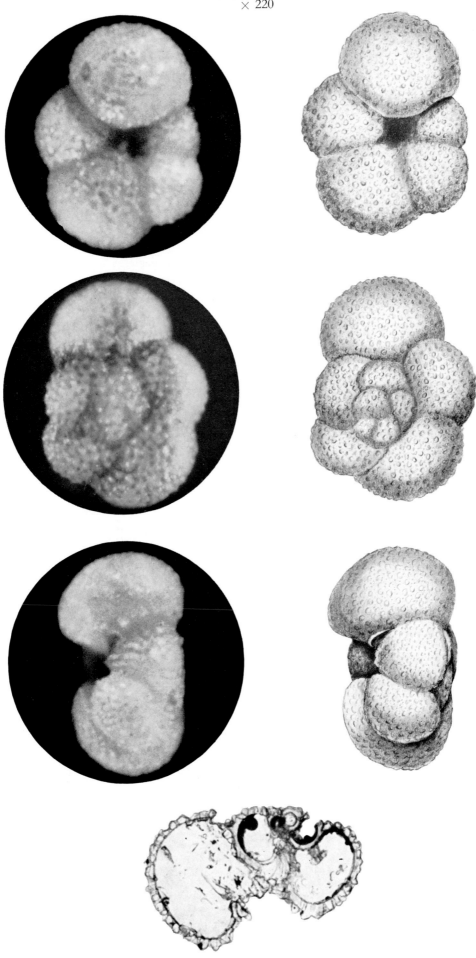

Globigerina daubjergensis BRONNIMANN

Reference   *Globigerina daubjergensis* BRONNIMANN, 1953: Note on planktonic Foraminifera from Danian localities of Jutland, Denmark. — Eclogae Geologicae Helvetiae, 45 (2):340, textfig. 1.

Type locality   Sample 38, from a quarry south-west of Stavnsbjerg farm, Daubjerg, Jutland, Denmark.

Diagnosis   Test medium to high trochospiral, biconvex, spiral side distinctly convex, umbilical side less so; equatorial periphery lobulate, axial periphery rounded.
Wall finely perforate, surface hispid.
Chambers inflated, globular or slightly compressed laterally, arranged in about three whorls; the 3½-4 chambers of the last whorl increase fairly rapidly in size.
Sutures on spiral side radial to slightly curved, depressed; on umbilical side almost radial, depressed.
Umbilicus narrow, shallow.
Aperture a small, interiomarginal, umbilical, subcircular arch, bordered by a faint lip.

Strat. distr.   Ranging throughout the *Globigerina daubjergensis* zone.

Remarks   The most distinct features of this very small species are the trochoid early chamber arrangement, the finely spinose wall surface and the small, shallow umbilicus. In stratigraphically younger forms a development of supplementary apertures may be present.
Locality of figured specimen is Station 23 (Plummer loc.), shallow ditch at road corner south-east of new Corsicana reservoir on the road to Mildred, Navarro County, Texas, U.S.A.

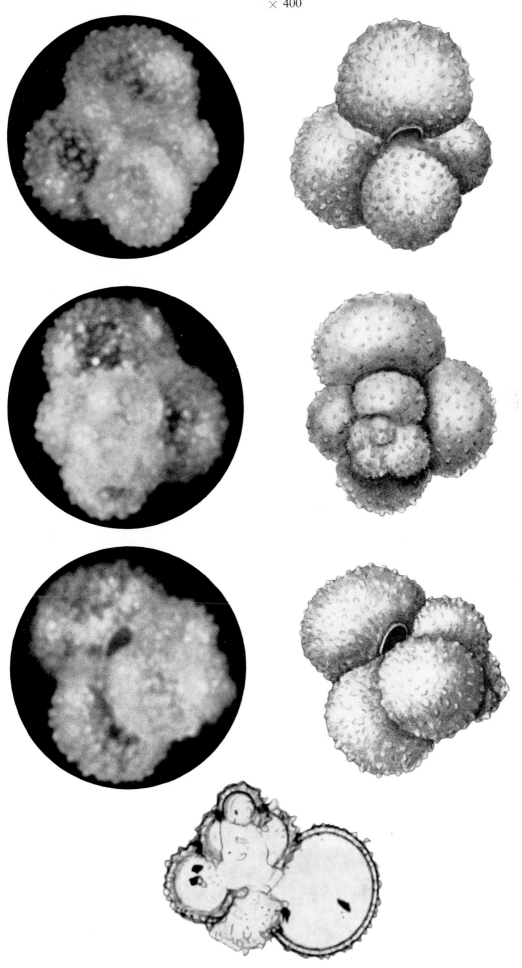

Globigerina daubjergensis
× 400

Globigerina gravelli BRONNIMANN

Reference   *Globigerina gravelli* BRONNIMANN, 1952: Trinidad Paleocene and Lower Eocene Globigerinidae. — Bulletins of American Paleontology, XXXIV (143):12, pl.1, figs. 16-18.

Type locality   Trinidad Leaseholds Ltd. sample No. 50506, from the ravine of the Ampelu River, Lizard Springs area, south-eastern Trinidad.

Diagnosis   Test low trochospiral, spiral side almost flat to slightly convex; umbilical side strongly convex; equatorial periphery slightly lobulate, axial periphery broadly rounded to sub-angular.
Wall perforate, surface distinctly nodose, in last chamber less so or smooth.
Chambers inflated, subglobular, compressed laterally, arranged in about three whorls; the five to six chambers of the last whorl increase slowly in size.
Sutures on spiral side oblique to slightly curved, depressed; on umbilical side radial, depressed.
Umbilicus fairly wide, deep.
Aperture a medium, interiomarginal, umbilical arch.

Strat. distr.   Base of *Globorotalia rex* zone to top of *Globorotalia formosa/aragonensis* zone.

Remarks   Locality of figured specimen is Tschopp 572, Cuba.

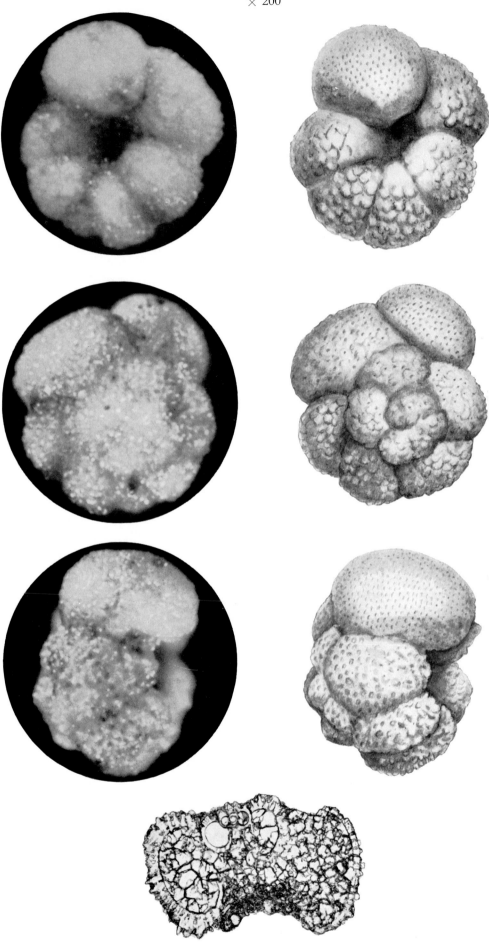

Globigerina ouachitaensis HOWE and WALLACE

Reference   *Globigerina ouachitaensis* HOWE and WALLACE, 1932: Foraminifera of the Jackson Eocene at Danville landing on the Ouachita, Catahoula Parish, Louisiana. — Louisiana Department of Conservation, Geol. Bull., 2:74, pl.10, fig. 7a-b.

Type locality   Danville landing on the Ouachita River, Catahoula Parish, Louisiana, U.S.A.

Diagnosis   Test small, medium to high trochospiral; equatorial periphery distinctly lobulate.
Wall finely perforate, rather delicate, surface very finely spinose, surface of last chamber smooth or almost smooth.
Chambers spherical, arranged in 2½-3 whorls; the four chambers of the last whorl increase moderately in size.
Sutures on spiral side slightly curved to radial, depressed, on umbilical side radial, depressed.
Umbilicus small.
Aperture a distinct, interiomarginal, umbilical arch, bordered by a rim.

Strat. distr.   Base of *Truncorotaloides rohri* zone to lower part of *Globigerina angulisuturalis* zone.

Remarks   *Globigerina parva* BOLLI, 1957, is considered a junior synonym.
Locality of figured specimen is the above-mentioned type locality.

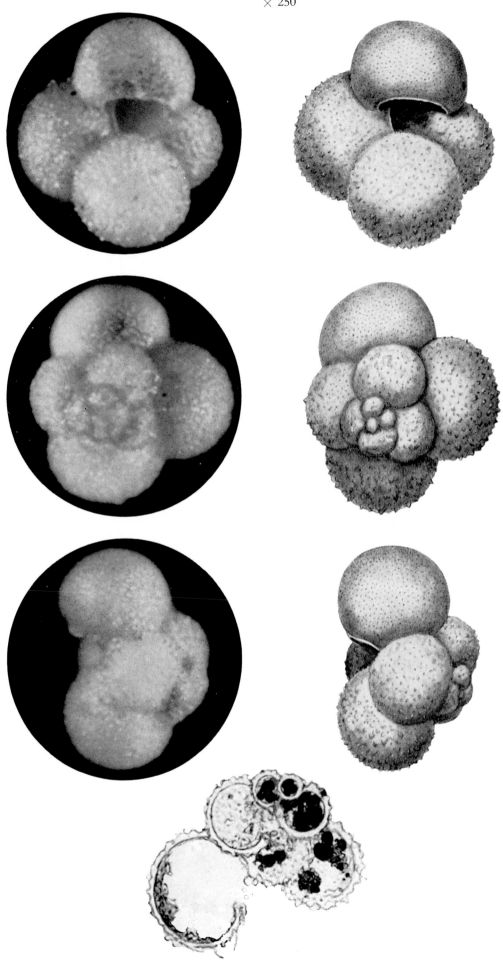

Globigerina primitiva (FINLAY)

Reference   *Globoquadrina primitiva* FINLAY, 1947: New Zealand Foraminifera: Key Species in Stratigraphy — No. 5 — New Zealand Journal of Science and Technology, 28 (5):291, pl.8, figs. 129-134.

Type locality   Sample number 517 B, Hampden, Otago Province, South Island, New Zealand.

Diagnosis   Test low trochospiral, spiral side slightly convex, umbilical side strongly inflated; equatorial periphery lobulate, axial periphery broadly rounded.
Wall perforate, surface distinctly spinose or nodose.
Chambers inflated, subglobular, compressed laterally, arranged in 2½ whorls; the three to four chambers of the last whorl increase rapidly in size.
Sutures on spiral side conspicuous, depressed; on umbilical side radial, depressed.
Umbilicus narrow.
Aperture a low, interiomarginal, umbilical arch, bordered usually by a rim. The aperture tends to become extraumbilical-umbilical in position.

Strat. distr.   Upper part of *Globorotalia pseudomenardii* zone to top of *Globorotalia formosa/aragonensis* zone.

Remarks   *Globigerina primitiva* (FINLAY) probably developed from *Globigerina velascoensis* CUSHMAN, from which it is distinguished mainly by its spinose surface.
Locality of figured specimen is the above-mentioned type locality.

Globigerina primitiva
× 220

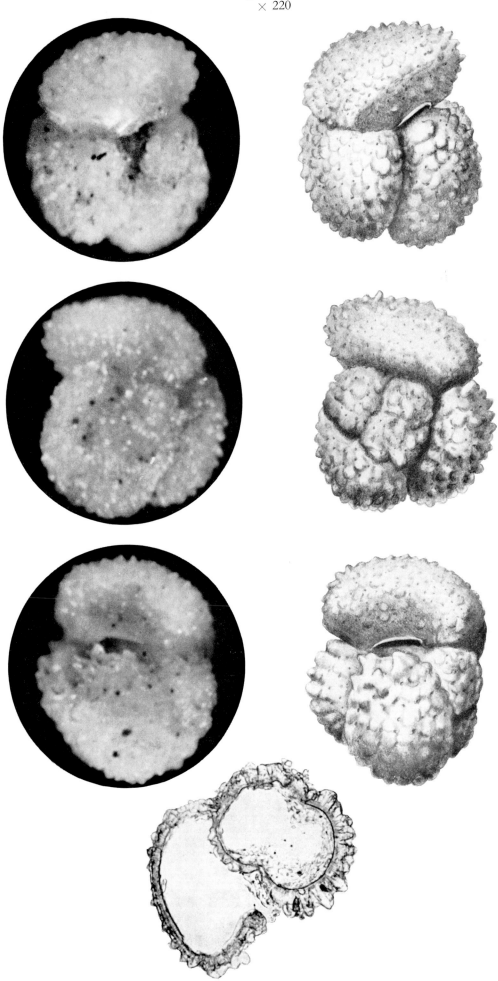

Globigerina senni (BECKMANN)

Reference *Sphaeroidinella senni* BECKMANN, 1953: Foraminiferen der Oceanic Formation (Eocaen-Oligocaen) von Barbados, Kl. Antillen. — Eclogae Geologicae Helvetiae, 46 (2):394, pl.26, figs. 2-4, textfig. 20.

Type locality Section on eastern slope of Mount Hillaby, at steep cliff on the left bank of upper Mount Hillaby River.

Diagnosis Test low trochospiral, spiral side slightly convex, umbilical side strongly convex, inflated; equatorial periphery slightly lobulate to almost circular, axial periphery very broadly rounded.
Wall coarsely perforate, surface distinctly ornamented with short spines or knobs and typically rugose at the umbilical edge.
Chambers inflated, in the young stage globular, in later stage subglobular, moderately compressed laterally, arranged in $2\frac{1}{2}$ whorls; the three to four chambers of the last whorl increase slowly in size; when there are four chambers in the last whorl, the last chamber is often reduced.
Sutures on spiral side slightly curved to radial, depressed; on umbilical side radial, depressed.
Umbilicus fairly narrow, deep.
Aperture an interiomarginal, umbilical arch.

Strat. distr. Base of *Globorotalia bullbrooki* zone to top of *Truncorotaloides rohri* zone.

Remarks Locality of figured specimen is DB 129, Trinidad.

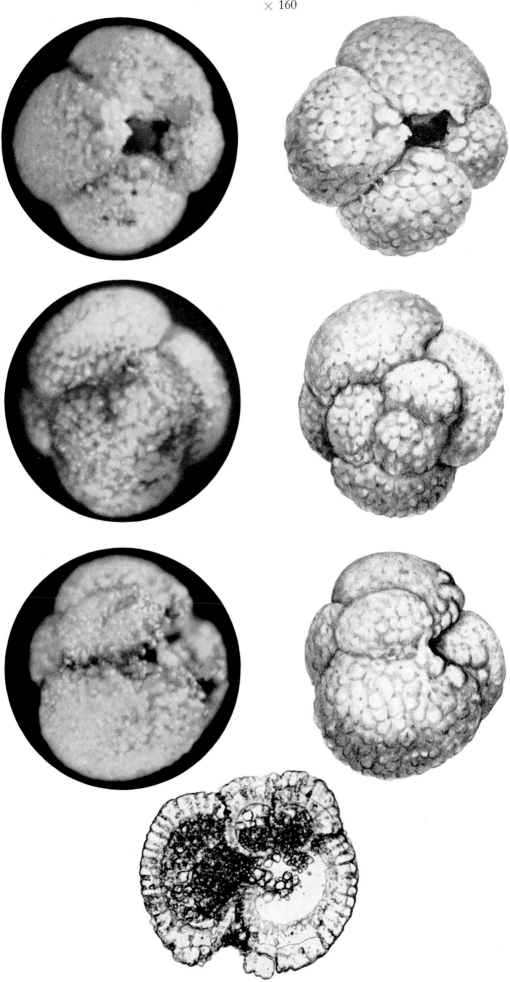

Globigerina soldadoensis BRONNIMANN

Reference   *Globigerina soldadoensis* BRONNIMANN, 1952: Trinidad Paleocene and Lower Eocene Globigerinidae. — Bulletins of American Paleontology, 34 (143) :9, pl.1, figs. 1-9.

Type locality   Trinidad Leaseholds Ltd. sample no. 50506, from the ravine of the Ampelu River, a small tributary of the Ortoire River system, Lizard Springs area, Trinidad.

Diagnosis   Test low trochospiral, spiral side almost flat to slightly convex, umbilical side strongly convex; equatorial periphery lobulate, axial periphery subangular to broadly rounded.
Wall coarsely perforate, surface covered with short blunt spines or knobs, distinctly so at the umbilical edge, surface of last chamber less so or smooth.
Chambers inflated, subglobular, compressed to strongly compressed laterally, arranged in 2½-3 whorls, the usually four chambers of the last whorl increase slowly in size.
Sutures on spiral side slightly curved to oblique, depressed; on umbilical side radial, depressed.
Umbilicus fairly narrow to fairly wide.
Aperture a low, interiomarginal, umbilical arch, bordered by a rim.
The aperture tends to become extraumbilical-umbilical in position.

Strat. distr.   Base of *Globorotalia velascoensis* zone to top of *Globorotalia formosa/aragonensis* zone.

Remarks   This species is apparently related to *Globigerina primitiva* (FINLAY), from which it is distinguished by its larger size and greater number of chambers. Occurring in the Uppermost Paleocene and throughout the Lower Eocene it is of limited stratigraphic significance.
Locality of figured specimen is HK 1831, Trinidad.

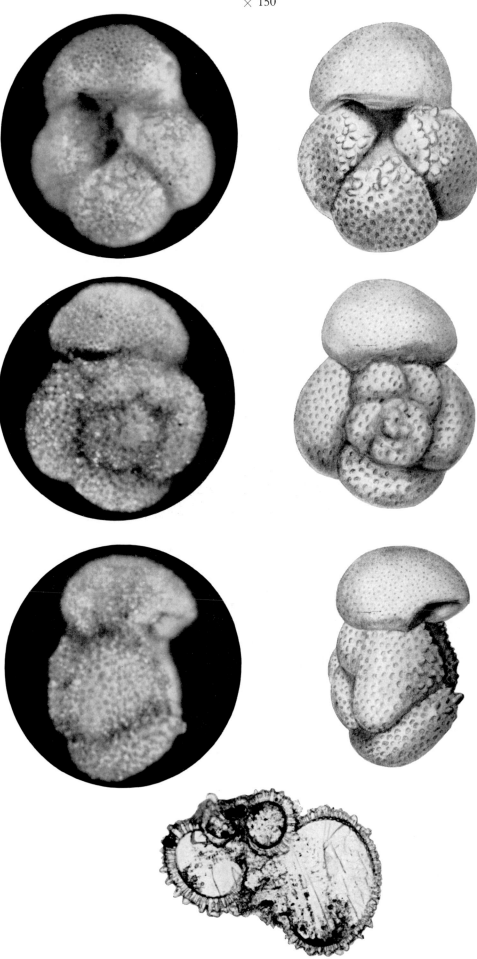

Globigerina triloculinoides PLUMMER

Reference *Globigerina triloculinoides* PLUMMER, 1926: Foraminifera of the Midway Formation in Texas. — University of Texas Bulletin, 2644:134, pl.VIII, fig. 10a-c.

Type locality Station 23, shallow ditch at road corner south-east of new Corsicana reservoir on road to Mildred, Navarro County, Texas, U.S.A.

Diagnosis Test low trochospiral, biconvex; equatorial periphery lobulate, axial periphery broadly rounded.
Wall coarsely perforate, surface distinctly reticulate.
Chambers inflated, globular to subglobular, arranged in two whorls; the 3½ chambers of the last whorl increase rapidly in size.
Sutures on spiral side curved, less so in the last chambers, depressed; on umbilical side almost radial, depressed.
Umbilicus shallow, narrow.
Aperture a low interiomarginal, umbilical arched slit, bordered by a distinct lip.

Strat. distr. Base of *Globigerina daubjergensis* zone to top of *Globorotalia angulata* zone.

Remarks The following species are closely related with intermediate forms and are thus of little stratigraphic value:

> *Globigerina pseudotriloba* WHITE, 1928
> *Globigerina triangularis* WHITE, 1928
> *Globigerina finlayi* BRONNIMANN, 1952
> *Globigerina hornibrooki* BRONNIMANN, 1952
> *Globigerina stainforthi* BRONNIMANN, 1952

Locality of figured specimen is the above-mentioned type locality.

Globigerina triloculinoides
× 160

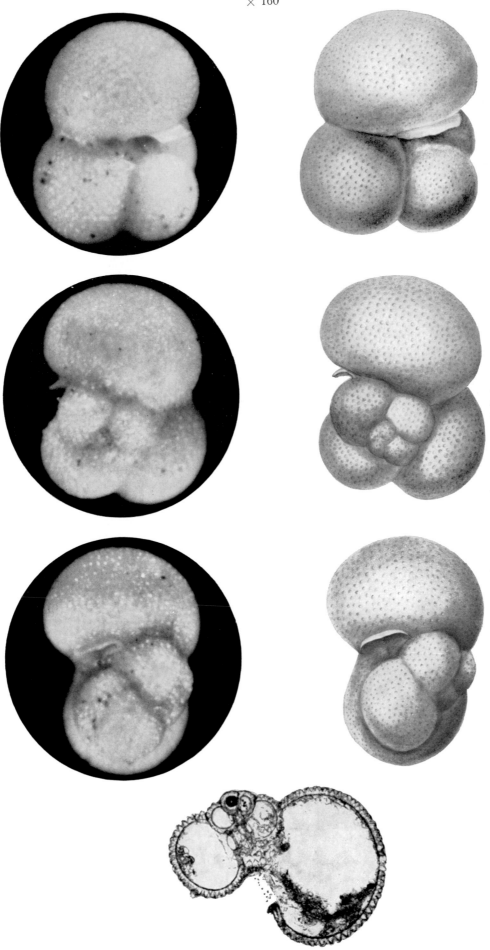

Globigerina yeguaensis WEINZIERL and APPLIN

Reference *Globigerina yeguaensis* WEINZIÈRL and APPLIN, 1929: The Claiborne Formation on the Coastal Domes. — Journal of Paleontology, 3 (4):408, pl.43, fig. 1a-b.

Type locality Rio Bravo Oil Co., Deussen B1, South Liberty, Liberty County, Texas, U.S.A.

Diagnosis Test low trochospiral, spiral side slightly convex, umbilical side convex; equatorial periphery distinctly lobulate, axial periphery broadly rounded.
Wall perforate, surface pitted and occasionally slightly spinose in early chambers, finely pitted in last chamber.
Chambers subspherical, somewhat compressed laterally, arranged in about three whorls; the three to four chambers of the last whorl increase rapidly to moderately in size.
Sutures on spiral side slightly curved to radial, depressed; on umbilical side radial, depressed.
Umbilicus fairly narrow.
Aperture a medium, interiomarginal, umbilical arch, bordered by a characteristic toothlike lip.

Strat. distr. Base of *Globigerapsis kugleri* zone to lower part of *Globigerina ampliapertura* zone.
Questionable occurrence in upper part of *Globigerina ampliapertura* zone.

Remarks Locality of figured specimen is E 227, Alabama, U.S.A.

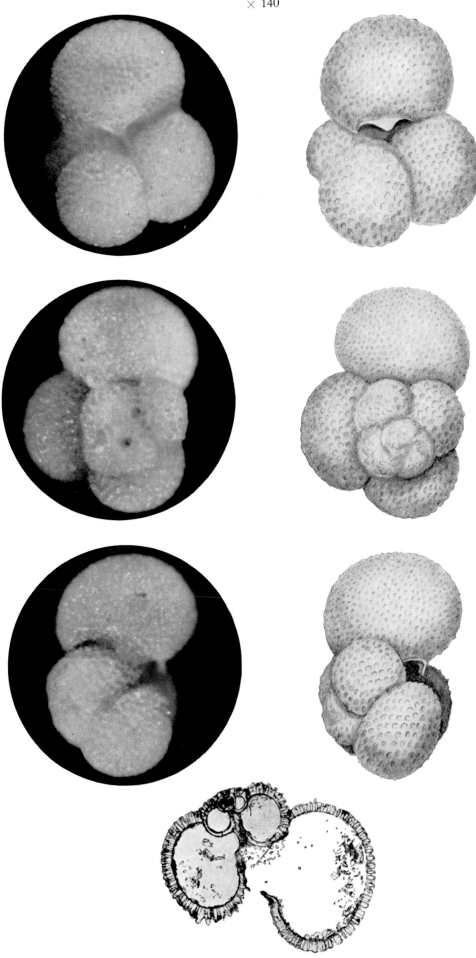

Globigerinatheka barri Brönnimann

Reference *Globigerinatheka barri* Brönnimann, 1952: *Globigerinoita* and *Globigerinatheka,* new genera from the Tertiary of Trinidad, B. W. I. — Contributions from the Cushman Foundation for Foraminiferal Research, 3 (1):27, textfig. 3a-c.

Type locality Trinidad Leaseholds Ltd. Cat. nos. 158007, 158028-158029, Mount Moriah Formation, Trinidad.

Diagnosis Test almost globular, early chambers arranged trochospirally.
Wall coarsely perforate, surface pitted.
Chambers of the early portion (*Globigerina* stage) globular, rapidly increasing in size; the large enveloping final chamber constitutes a considerable part of the test
Sutures slightly curved to radial, depressed.
Primary aperture is covered by the final enveloping chamber, the secondary sutural apertures along the margin of the final chamber are covered by small bullae, each of which has one or more small arched infralaminal accessory apertures.

Strat. distr. Base of *Globorotalia lehneri* zone to top of *Globigerapsis mexicana* zone. Questionable occurrence in lower part of the *Globorotalia cerroazulensis* zone.

Remarks *Globigerinatheka barri* Brönnimann developed from *Globigerapsis kugleri* Bolli, Loeblich and Tappan, from which it differs in possessing sutural bullae. Locality of figured specimen is DB 276, Trinidad.

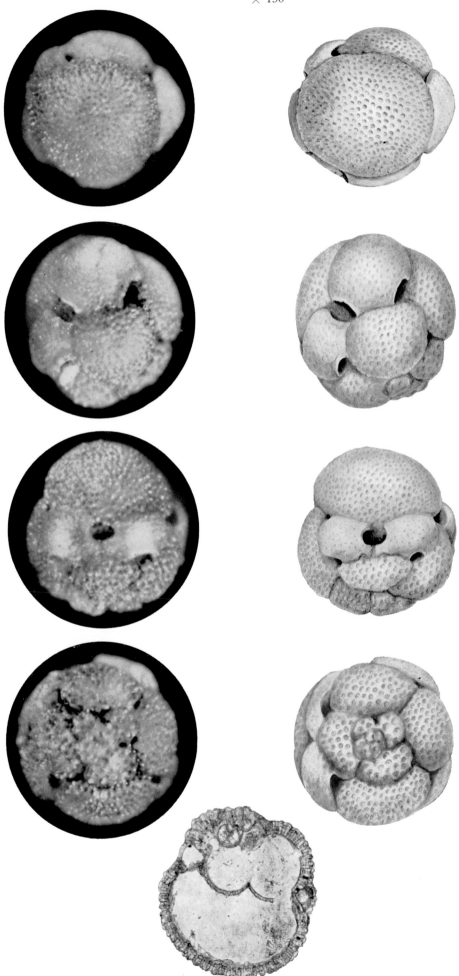

Globorotalia abundocamerata BOLLI

Reference   *Globorotalia angulata abundocamerata* BOLLI, 1957: The Genera *Globigerina* and *Globorotalia* in the Paleocene-Lower Eocene Lizard Springs Formation of Trinidad, B. W. I. — United States National Museum Bulletin, 215:74, pl.17, figs. 4-6.

Type locality   Trinidad Leaseholds, Ltd., well Guayaguayare 159, Trinidad, depth 4524-4536 feet.

Diagnosis   Test very low trochospiral, spiral side almost flat, inner whorl occasionally slightly raised, umbilical side strongly convex; equatorial periphery slightly lobulate, almost circular, axial periphery acute to subacute.
Wall perforate, surface finely spinose to slightly nodose.
Chambers subangular, inflated, arranged in about $2\frac{1}{2}$ whorls; the six to seven chambers of the last whorl increase slowly in size.
Sutures on spiral side curved, slightly depressed; on umbilical side radial, depressed.
Umbilicus fairly narrow to fairly wide, deep.
Aperture an interiomarginal, extraumbilical-umbilical narrow slit.

Strat. distr.   Ranging throughout the *Globorotalia angulata* zone and the lower part of the *Globorotalia pseudomenardii* zone.

Remarks   *Globorotalia abundocamerata* BOLLI is a multichambered form related to *Globorotalia angulata* (WHITE).
Locality of figured specimen is well San Roman I, Guatemala, depth 1870-1880 feet.

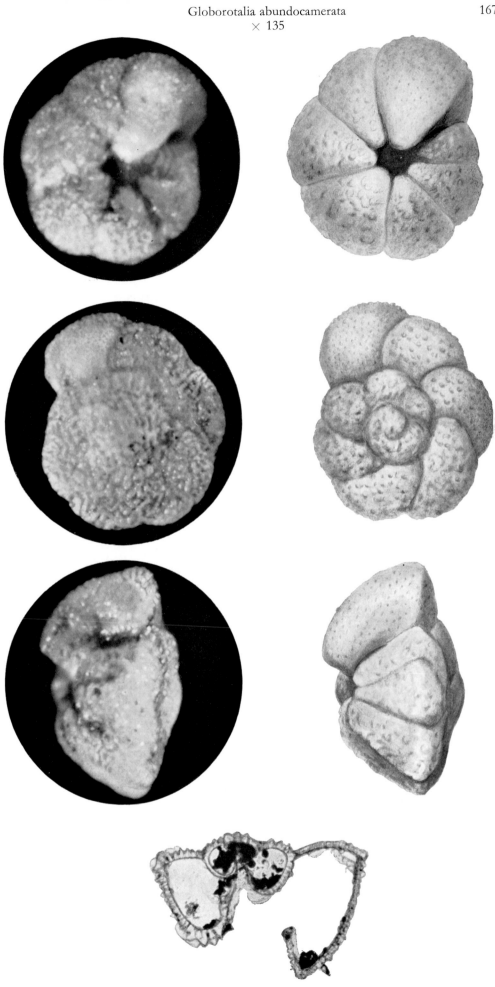

Globorotalia aequa CUSHMAN and RENZ

Reference   *Globorotalia crassata* var. *aequa* CUSHMAN and RENZ, 1942: Eocene, Midway, Foraminifera from Soldado Rock, Trinidad. — Contributions from the Cushman Laboratory for Foraminiferal Research, 18:12, pl.3a-c.

Type locality   Midway Eocene, Soldado Formation, Soldado Rock, Trinidad.

Diagnosis   Test very low trochospiral, spiral side flat to slightly convex, umbilical side strongly convex; equatorial periphery lobulate, axial periphery acute, a faint keel occasionally ornamented with spines or small knobs.

Wall perforate, surface of early chambers spinose or nodose, that of last chamber usually smooth.

Chambers angular, inflated, arranged in 2½ whorls; the chambers of the last whorl, usually four in number, increase rapidly in size, the last chamber forming a considerable part of the surface of the test.

Sutures on spiral side strongly curved, slightly depressed; on umbilical side radial, distinctly depressed.

Umbilicus narrow, deep.

Aperture a low, interiomarginal, extraumbilical-umbilical arch, bordered by a faint lip.

Strat. distr.   Base of *Globorotalia pseudomenardii* zone to top of *Globorotalia velascoensis* zone. Questionable occurrence in the *Globorotalia rex* zone.

Remarks   *Globorotalia lacerti* CUSHMAN and RENZ is a junior synonym of *Globorotalia aequa*. Locality of figured specimen is HK 1831, Trinidad.

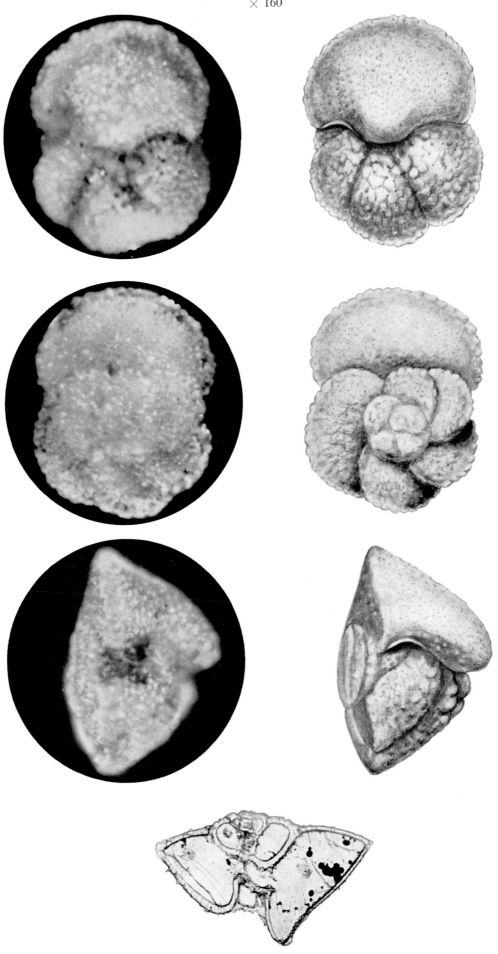

Globorotalia angulata (WHITE)

Reference   *Globigerina angulata* WHITE, 1928: Some index Foraminifera of the Tampico embayment area of Mexico. — Journal of Paleontology, 2 (3):191, pl.27, fig. 13a-c.

Type locality   Columbus Station of the Tampico Monterey railroad line, Tampico embayment, Mexico.

Diagnosis   Test very low trochospiral, spiral side almost flat, umbilical side strongly convex; equatorial periphery lobulate, a faint keel may be present, axial periphery acute to subacute.
Wall perforate, surface spinose, in early chambers sometimes slightly nodose.
Chambers angular to subangular, inflated, arranged in about 2½ whorls; the chambers of the last whorl, usually five in number, increase fairly slowly in size.
Sutures on spiral side curved, depressed; on umbilical side radial, strongly depressed.
Umbilicus fairly narrow, deep.
Aperture interiomarginal, extraumbilical-umbilical, a narrow slit, bordered by a faint lip.

Strat. distr.   Ranging throughout the *Globorotalia angulata* zone. Questionable occurrence in upper part of *Globorotalia uncinata* zone.

Remarks   This species belongs to the evolutionary sequence of *Globorotalia uncinata* BOLLI — *Globorotalia angulata* (WHITE) — *Globorotalia aequa* CUSHMAN and RENZ.
Locality of figured specimen is H.178, Turkey.

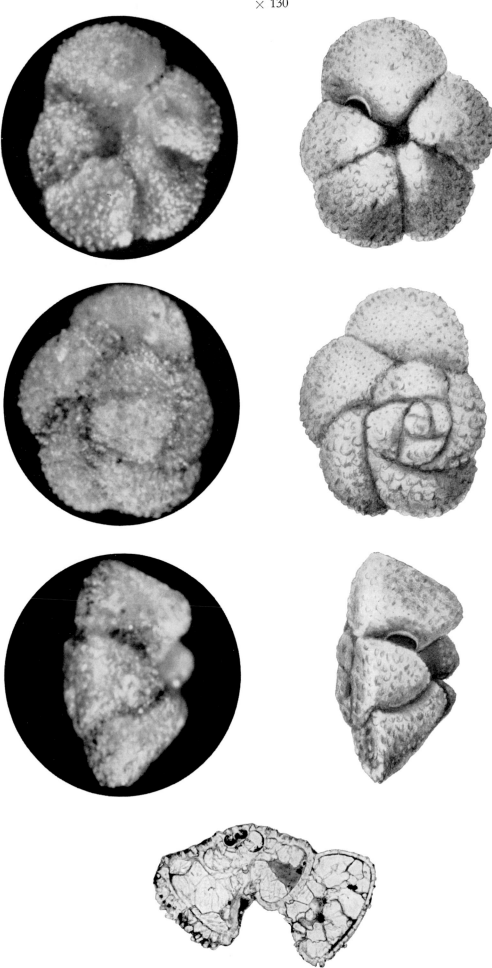

Globorotalia aragonensis Nuttall

Reference   *Globorotalia aragonensis* Nuttall, 1930: Eocene Foraminifera from Mexico. — Journal of Paleontology, 4 (3):288, pl.24, figs. 6-8, 10-11.

Type locality   La Antigua, Rio La Puerta, Tampico region, Mexico.

Diagnosis   Test very low trochospiral, spiral side flat to slightly convex, umbilical side strongly convex, inflated; equatorial periphery nearly circular, axial periphery angular with peripheral keel, ornamented with short spines or knobs.
Wall finely perforate, surface of early chambers of last whorl rugose, that of last chambers smooth.
Chambers angular, inflated, arranged in about three whorls; the six to seven chambers of the last whorl increase slowly in size.
Sutures on spiral side curved, somewhat raised and often slightly beaded; on umbilical side radial, depressed.
Umbilicus narrow, deep.
Aperture an interiomarginal, extraumbilical-umbilical arch, bordered by a faint lip.

Strat. distr.   Base of *Globorotalia formosa/aragonensis* zone to top of *Globigerapsis kugleri* zone.

Remarks   *Globorotalia aragonensis* Nuttall differs from *Globorotalia formosa* Bolli in having a less lobate periphery and a more closed umbilicus.
Locality of figured specimen is the above-mentioned type locality.

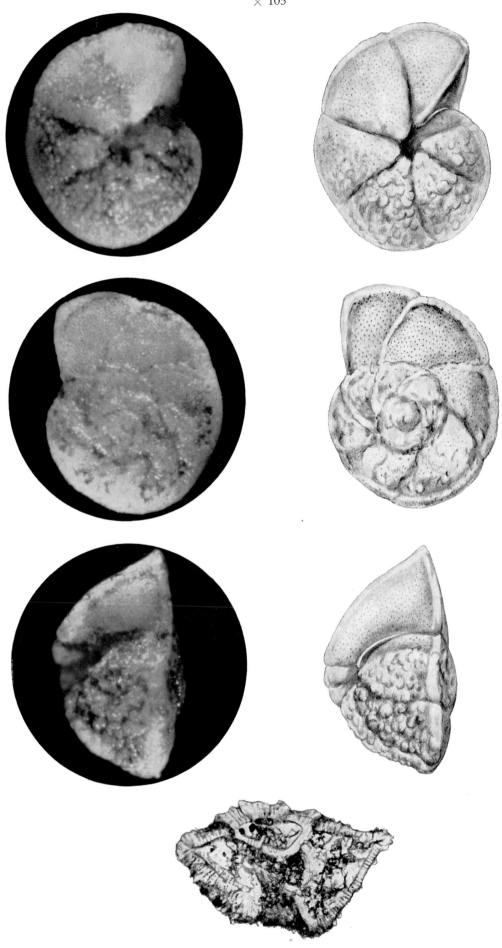

Globorotalia aspensis (COLOM)

Reference   *Globigerina aspensis* COLOM, 1954: Estudio de las biozonas con foraminíferos del Terciario de Alicante. — Bol. Inst.Geol. y Min. España, 66:151, pl.3, figs. 1-35, pl.4, figs. 1-31.

Type locality   3 km along the Aspe-Crevillente road, in the south-eastern part of the province of Alicante, Spain.

Diagnosis   Test very low trochospiral, spiral side usually slightly concave, umbilical side convex; equatorial periphery slightly lobulate, axial periphery broadly rounded.
Wall perforate, surface covered with short spines or knobs.
Chambers inflated, subglobular, somewhat compressed laterally, arranged in about three whorls, the initial whorls small compared with the last one; the six to seven chambers of the last whorl barely increase in size.
Sutures on spiral side slightly curved to slightly oblique, depressed; on umbilical side radial, depressed.
Umbilicus fairly narrow.
Aperture a low, interiomarginal, extraumbilical-umbilical arch, bordered by a rim.

Strat. distr.   Upper part of *Globorotalia formosa/aragonensis* zone to lower part of *Globigerapsis kugleri* zone. Questionable occurrences in middle part of *Globorotalia formosa/aragonensis* zone and upper part of *Globigerapsis kugleri* zone.

Remarks   *Globorotalia globigeriniformis* VAN BELLEN is most probably synonymous with the species described above.
Locality of figured specimen is Tschopp 3181, Cuba.

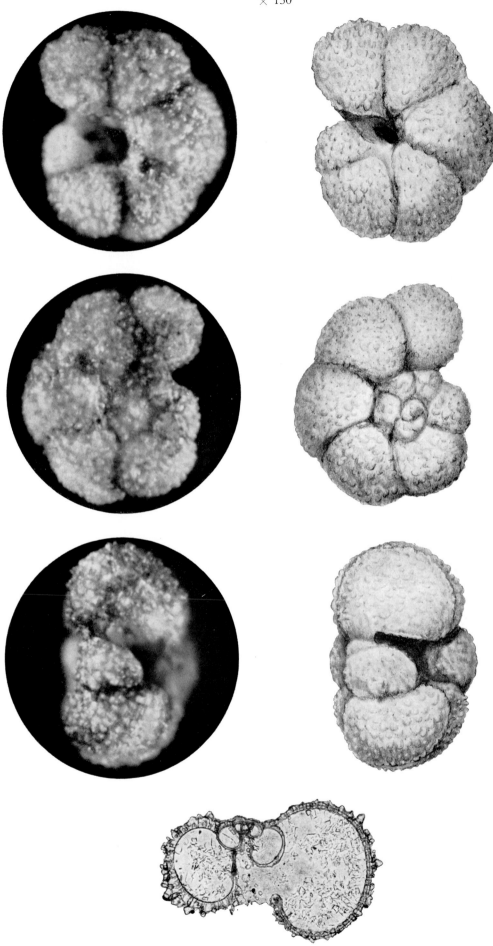

Globorotalia bolivariana (PETTERS)

Reference *Globigerina wilsoni* COLE subsp. *bolivariana* PETTERS, 1954: Tertiary and Upper Cretaceous Foraminifera from Columbia S.A. — Contributions from the Cushman Foundation for Foraminiferal Research, 5 (1):39, pl.8, fig. 9a-c.

Type locality In a 50° south-east-dipping limb of a small syncline, 2230 metres west-northwest of Carreto, on the road to San Cayetano, Department of Bolivar, Colombia.

Diagnosis Test very low trochospiral, tending to become involute on spiral side; equatorial periphery slightly lobulate, axial periphery broadly rounded.
Wall finely perforate, surface of last chamber smooth, surface of earlier chambers with short spines or nodose.
Chambers inflated, subglobular, arranged in 2½ whorls; the four chambers of the last whorl increase rapidly in size.
Sutures on spiral side slightly curved, depressed; on umbilical side radial, depressed. Umbilicus narrow.
Aperture a narrow, interiomarginal, extraumbilical-umbilical slit, often extending towards the spiral side, bordered by a pronounced lip.

Strat. distr. Base of *Globorotalia bullbrooki* zone to top of *Truncorotaloides rohri* zone.

Remarks Locality of figured specimen is DB 272, Trinidad.

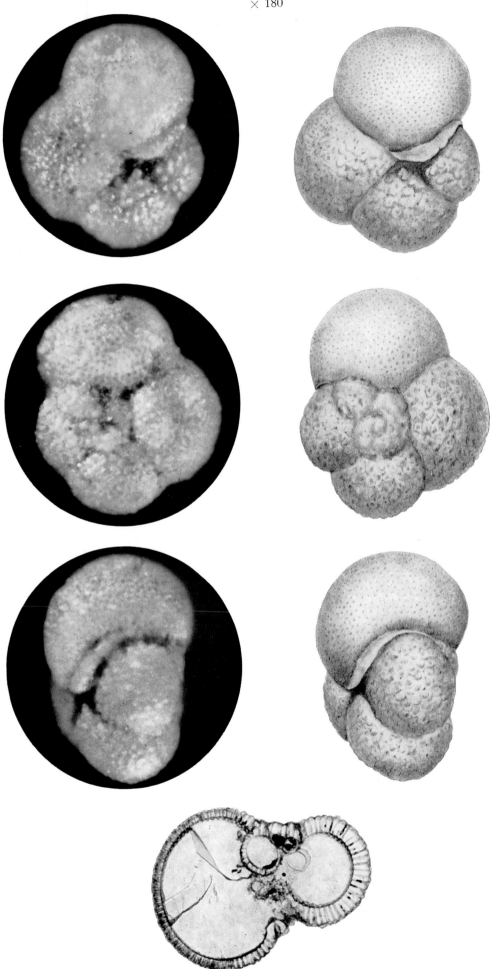

Globorotalia broedermanni CUSHMAN and BERMUDEZ

Reference *Globorotalia broedermanni* CUSHMAN and BERMUDEZ, 1949: Some Cuban species of *Globorotalia*. — Contributions from the Cushman Laboratory for Foraminiferal Research, 25:40, pl.7, figs. 22-24.

Type locality Bermudez station 349, 500 m south of Capdevila, Habana Province, Cuba.

Diagnosis Test low trochospiral, biconvex, moderately compressed; equatorial periphery nearly circular, axial periphery rounded to subangular.
Wall coarsely perforate, surface covered with short spines.
Chambers subangular to subglobular, inflated, arranged in about 2½ whorls; the chambers of the last whorl, usually six in number, increase slowly in size.
Sutures on spiral side slightly curved, slightly depressed; on umbilical side radial, slightly depressed.
Umbilicus narrow.
Aperture a low, interiomarginal, extraumbilical-umbilical arch.

Strat. distr. Base of *Globorotalia formosa/aragonensis* zone to top of *Globigerapsis kugleri* zone. *Globorotalia* cf. *broedermanni* occurs throughout the *Globorotalia rex* zone.

Remarks *Globorotalia broedermanni* CUSHMAN and BERMUDEZ and *Globorotalia convexa* SUBBOTINA are probably closely related or even synonymous.
Locality of figured specimen is Tschopp 3181, Cuba.

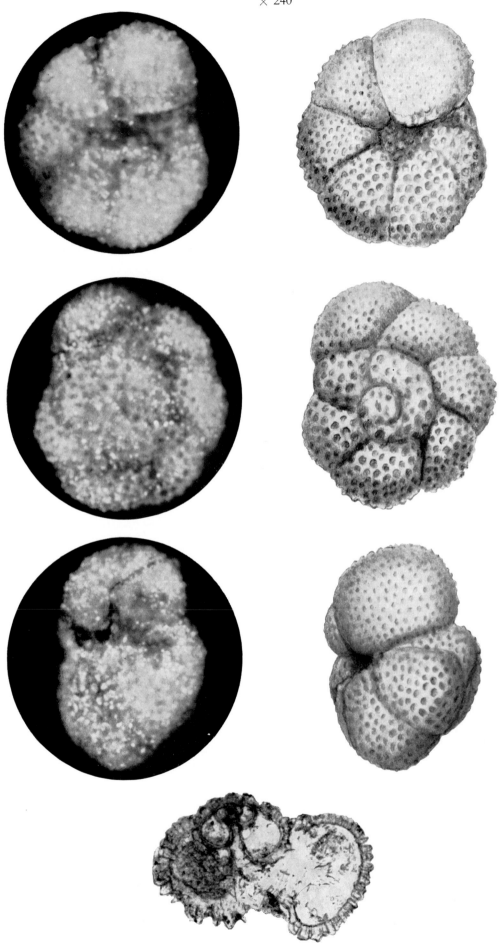

Globorotalia bullbrooki Bolli

Reference *Globorotalia bullbrooki* Bolli, 1957: Planktonic Foraminifera from the Eocene Navet and San Fernando Formations of Trinidad, B.W.I. — United States National Museum Bulletin, 215 : 167, pl.38, figs. 4a-5c.

Type locality Holotype from an outcrop on the left side of the right branch of the Nariva River, about 450 feet from its junction, Central Range, Trinidad. Navet Formation.

Diagnosis Test low trochospiral, spiral side almost flat, umbilical side strongly convex; equatorial periphery lobulate, axial periphery subangular.
Wall perforate, surface covered with short spines or knobs.
Chambers subangular, inflated, arranged in about $2\frac{1}{2}$ whorls; the 4-$4\frac{1}{2}$ chambers of the last whorl increase rapidly in size.
Sutures on spiral side oblique to slightly curved, slightly depressed; on umbilical side radial, depressed.
Umbilicus narrow, deep.
Aperture an interiomarginal, extraumbilical-umbilical arch.

Strat. distr. Ranging throughout the *Globorotalia bullbrooki* zone and the lower part of the *Globigerapsis kugleri* zone. Questionable occurrence in the upper part of the *Globigerapsis kugleri* zone.

Remarks *Globorotalia bullbrooki* Bolli may be related to species like *Globorotalia crassata* (Cushman) and *Globorotalia spinuloinflata* (Bandy).
Locality of figured specimen is Tschopp 372, Cuba.

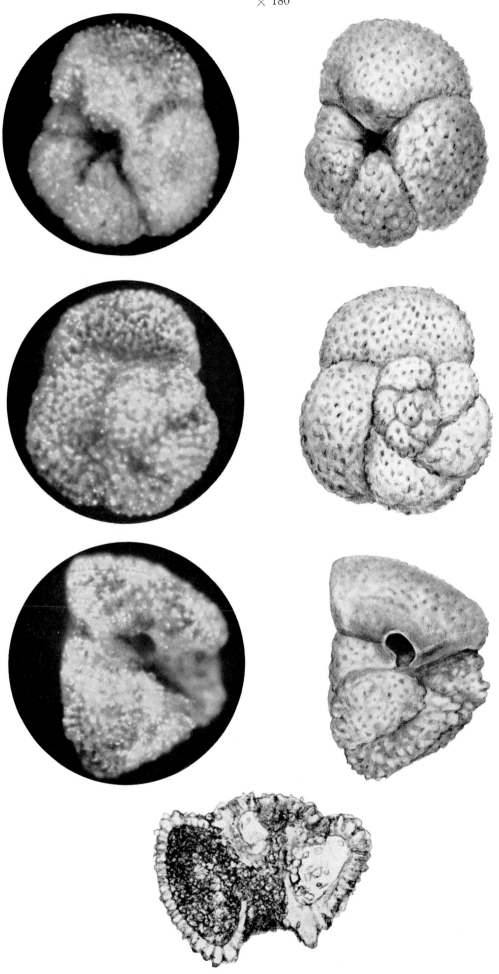

Globorotalia centralis Cushman and Bermudez

Reference   *Globorotalia centralis* Cushman and Bermudez, 1937: Further new species of Foraminifera from the Eocene of Cuba. — Contributions from the Cushman Laboratory for Foraminiferal Research, 13 (1):26, pl.2, figs. 62-65.

Type locality   Bermudez Station 92, under R.R. bridge on Central Highway located in Jicotea, Santa Clara Province, Cuba.

Diagnosis   Test low trochospiral, spiral side almost flat to slightly convex, umbilical side distinctly convex; equatorial periphery slightly lobulate to almost circular, axial periphery rounded to subangular.
Wall distinctly perforate, surface smooth.
Chambers inflated, compressed laterally, arranged in $3-3\frac{1}{2}$ whorls; the four to five chambers of the last whorl increase fairly rapidly in size.
Sutures on spiral side slightly curved to oblique, slightly depressed; on umbilical side radial, depressed.
Umbilicus narrow.
Aperture a fairly large, elongate arch, bordered by a narrow rim, interiomarginal, extraumbilical-umbilical.

Strat. distr.   Upper part of *Globigerapsis kugleri* zone to top of *Globorotalia cerroazulensis* zone.

Remarks   There is some variation in chamber shape from rounded to subangular. The more subangular specimens may be regarded as transitional to *Globorotalia cerroazulensis* (Cole). Locality of figured specimen is well Criollo 1, 1895[1], Cuba.

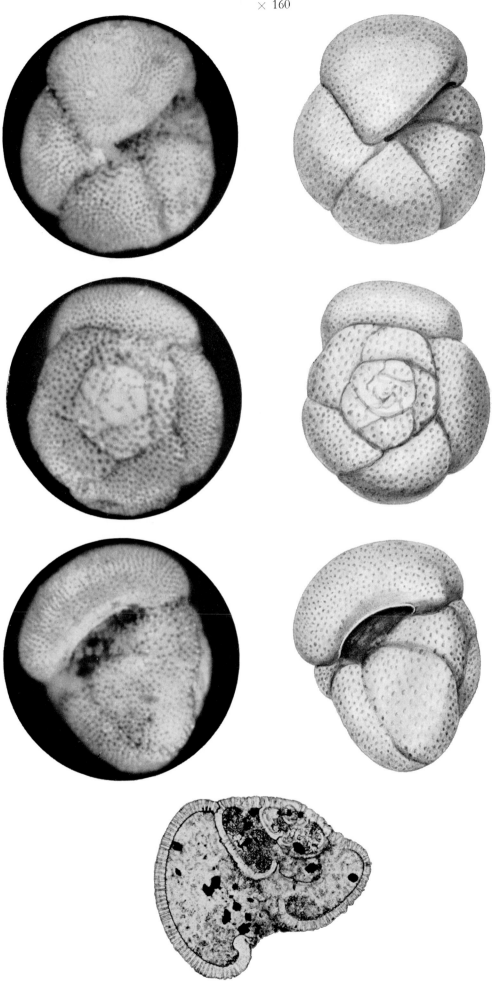

Globorotalia cerroazulensis (COLE)

Reference *Globigerina cerro-azulensis* COLE, 1928: A foraminiferal fauna from the Chapapote Formation in Mexico. — Bulletins of American Paleontology, XIV (53):217, pl. 32, figs. 11-13.

Type locality On the Tuxpan River near the village of Chapapote, Chapapote Formation, Mexico.

Diagnosis Test low trochospiral, spiral side almost flat, inner portion may be slightly convex, umbilical side distinctly convex; equatorial periphery slightly lobulate to nearly circular, axial periphery acute to subacute with a faint keel which may be absent in the last chamber.

Wall finely perforate, surface smooth except for surface of first chambers of last whorl on umbilical side, which may be slightly spinose.

Chambers inflated, strongly compressed laterally, the last chamber less so, arranged in 3-3½ whorls; the four chambers of the last whorl increase rapidly in size, the last chamber constitutes a considerable part of the surface of the test.

Sutures on spiral side distinctly curved, slightly raised; on umbilical side radial, depressed. Umbilicus narrow.

Aperture a fairly large, elongate arch, bordered by a narrow rim, interiomarginal, extraumbilical-umbilical.

Strat. distr. Base of *Globigerapsis mexicana* zone to top of *Globorotalia cerroazulensis* zone.

Remarks *Globorotalia cocoaensis* CUSHMAN is a synonym of the species, described above. Locality of figured specimen is E 227, Alabama, U.S.A.

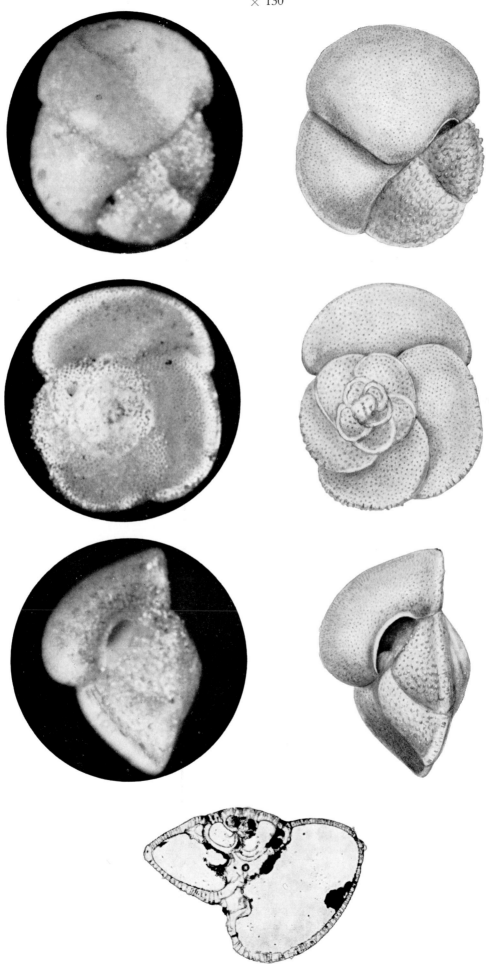

Globorotalia compressa (Plummer)

Reference	*Globigerina compressa* Plummer, 1926: Foraminifera of the Midway Formation in Texas. — University of Texas Bulletin, 2644:135, pl. VIII, fig. 11a-c.
Type locality	Station 23, shallow ditch at road corner south-east of new Corsicana reservoir on the road to Mildred, Navarro County, Texas, U.S.A.
Diagnosis	Test very low trochospiral, inflated; equatorial periphery distinctly lobulate, axial periphery subrounded to subacute. Wall finely perforate, surface smooth. Early chambers inflated, subglobular, later chambers moderately inflated, ovate, arranged in about 2½ whorls; the commonly five chambers of the last whorl increase fairly rapidly in size. Sutures on spiral side curved, less so in the last chambers, depressed; on umbilical side almost radial, depressed. Umbilicus fairly wide, open. Aperture interiomarginal, extraumbilical-umbilical, an arched opening extending nearly to the periphery, bordered by a distinct lip.
Strat. distr.	Base of *Globigerina daubjergensis* zone to top of *Globorotalia angulata* zone.
Remarks	*Globorotalia compressa* (Plummer) is the ancestor of *Globorotalia ehrenbergi* Bolli, from which it is distinguished by less compressed chambers and absence of a peripheral keel. Locality of figured specimen is the above-mentioned type locality.

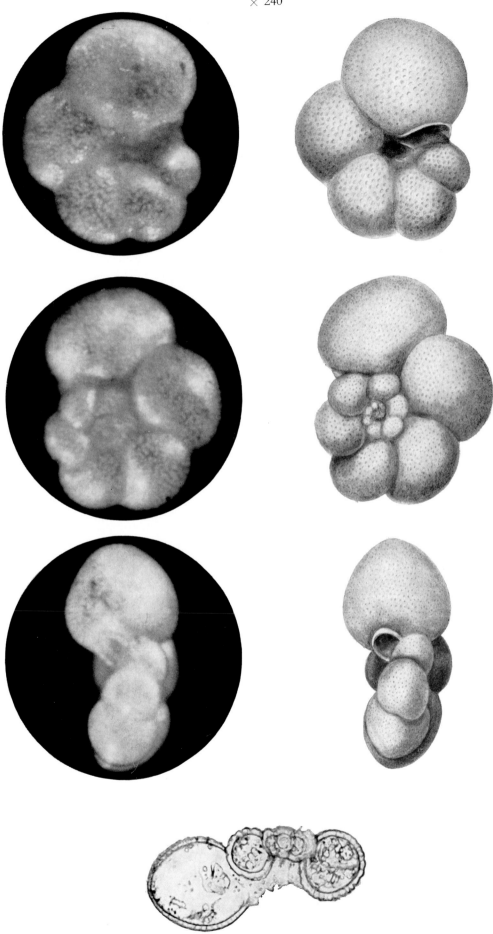

Globorotalia ehrenbergi BOLLI

Reference *Globorotalia ehrenbergi* BOLLI, 1957: The Genera *Globigerina* and *Globorotalia* in the Paleocene-Lower Eocene Lizard Springs Formation of Trinidad, B.W.I. — United States National Museum Bulletin, 215:77, pl. 20, figs. 18-20.

Type locality Trinidad Leaseholds, Ltd., well Guayaguayare 159, Trinidad, depth 4524-4536 feet.

Diagnosis Test low trochospiral, compressed; equatorial periphery distinctly lobulate, axial periphery subacute to acute, last chamber often with a faint pseudokeel.
Wall finely perforate, surface smooth.
Chambers ovate, becoming gradually more compressed, arranged in $2\frac{1}{2}$-3 whorls; the five chambers of the last whorl increase fairly rapidly in size.
Sutures on spiral side slightly curved, depressed; on umbilical side radial, depressed.
Umbilicus fairly wide, shallow.
Aperture interiomarginal, extraumbilical-umbilical, a low arch, bordered by a lip.

Strat. distr. Ranging throughout the *Globorotalia angulata* zone and the lower part of the *Globorotalia pseudomenardii* zone.

Remarks *Globorotalia ehrenbergi* BOLLI developed from *Globorotalia compressa* (PLUMMER) and is the ancestor of *Globorotalia pseudomenardii* BOLLI.
Locality of figured specimen is Dyr el Kef, W. Tunisia.

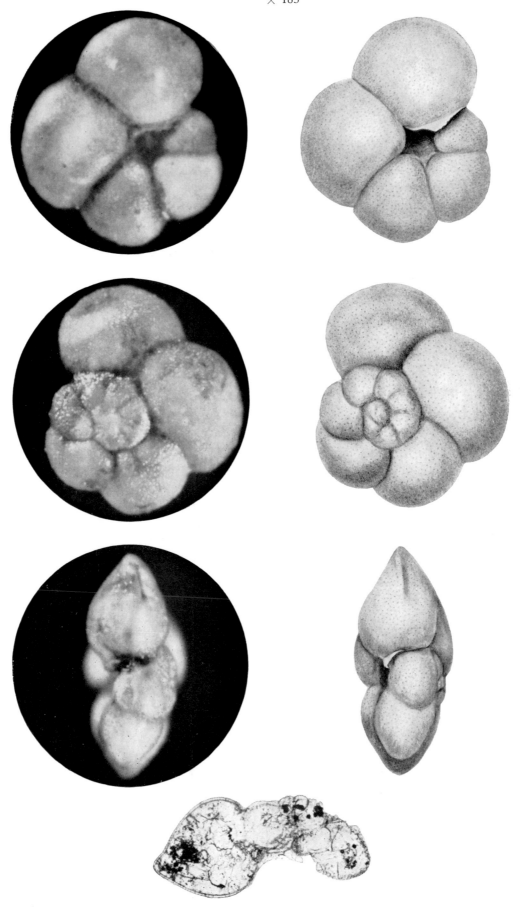

Globorotalia formosa BOLLI

Reference   *Globorotalia formosa formosa* BOLLI, 1957: The Genera *Globigerina* and *Globorotalia* in the Paleocene-Lower Eocene Lizard Springs Formation of Trinidad, B.W.I. — United States National Museum Bulletin, 215:76, pl. 18, figs. 1-3.

Type locality   Ampelu Ravine, about 1¼ miles south-east of the junction between the Rio Claro-Guayaguayare road and the Trinidad Central Oilfields road leading to the abandoned Lizard Springs oilfield, Trinidad.

Diagnosis   Test very low trochospiral, spiral side almost flat to slightly convex, umbilical side strongly convex; equatorial periphery slightly lobulate, axial periphery angular with pronounced keel, ornamented with short spines or knobs.
Wall finely perforate, surface of early chambers of last whorl rugose, especially on the umbilical side.
Chambers angular, inflated, arranged in about three whorls; the six to eight chambers of the last whorl increase slowly in size.
Sutures on spiral side curved, raised and beaded; on umbilical side radial, depressed.
Umbilicus fairly wide, deep, open.
Aperture an interiomarginal, extraumbilical-umbilical arch, bordered by a faint lip or rim.

Strat. distr.   Ranging throughout the *Globorotalia formosa/aragonensis* zone.

Remarks   *Globorotalia formosa* BOLLI differs from *Globorotalia gracilis* BOLLI in its more robust test, larger size and greater number of chambers in the last whorl. It differs from *Globorotalia aragonensis* NUTTALL in its more lobulate periphery and wider umbilicus. Locality of figured specimen is Tschopp 72, Cuba.

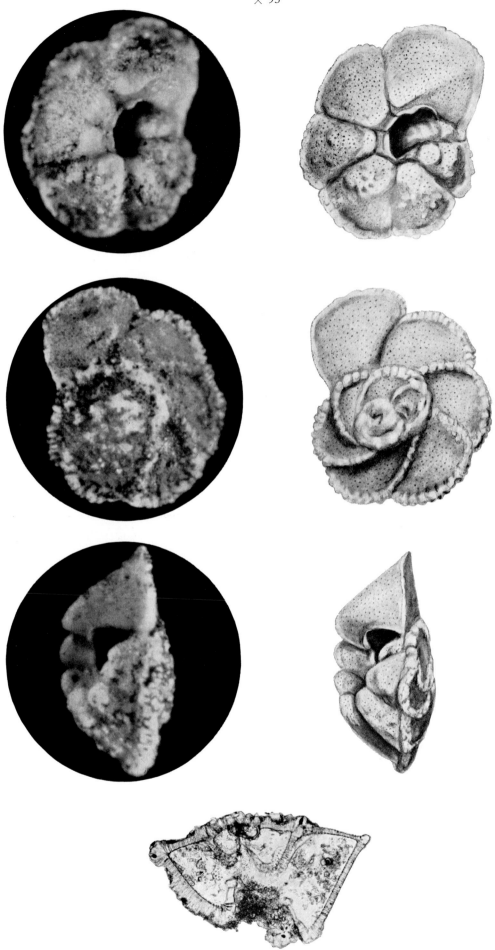

Globorotalia gracilis BOLLI

Reference   *Globorotalia formosa gracilis* BOLLI, 1957: The Genera *Globigerina* and *Globorotalia* in the Paleocene-Lower Eocene Lizard Springs Formation of Trinidad, B.W.I. — United States National Museum Bulletin, 215:75, pl. 18, figs. 4-6.

Type locality   Trinidad Leaseholds, Ltd., well Guayaguayare 159, Trinidad, depth 3707-3713 feet.

Diagnosis   Test very low trochospiral, spiral side almost flat or slightly convex, umbilical side markedly convex; equatorial periphery lobulate, axial periphery angular with a keel ornamented with spines or knobs.

Wall perforate, surface spinose or nodose, in last chamber less so. Chambers angular, inflated, arranged in $2\frac{1}{2}$-3 whorls; the five to six chambers of the last whorl increase fairly rapidly in size.

Sutures on spiral side curved, raised, ornamented with spines or small knobs; on umbilical side radial, depressed.

Umbilicus fairly narrow to fairly wide, deep.

Aperture a low, interiomarginal, extraumbilical-umbilical arch, bordered by a faint lip or rim.

Strat. distr.   Ranging throughout the *Globorotalia rex* zone and the lower part of the *Globorotalia formosa/aragonensis* zone.

Remarks   Locality of figured specimen is HK 1831, Trinidad.

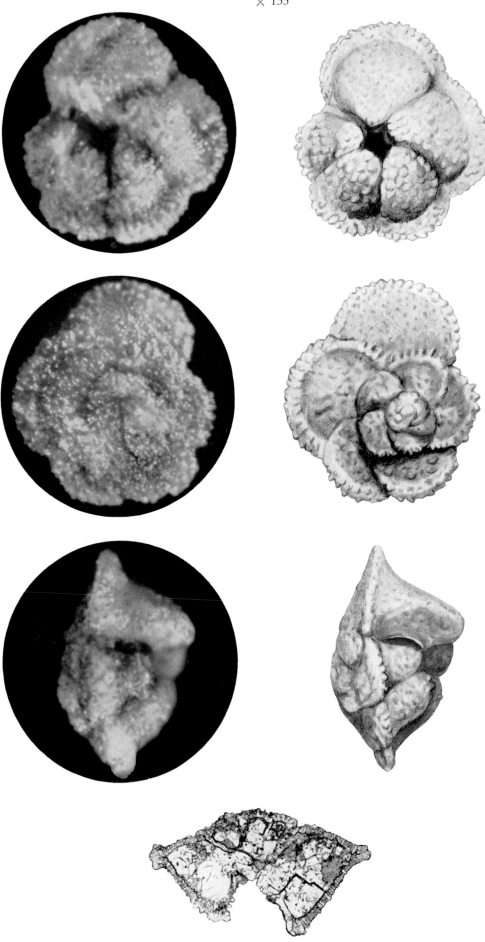

Globorotalia increbescens (BANDY)

Reference *Globigerina increbescens* BANDY, 1949: Eocene and Oligocene Foraminifera from Little Stave Creek, Clarke County, Alabama. — Bulletins of American Paleontology, 32 (131):120, pl. 23, fig. 3.

Type locality Upper part Jackson Formation; about 2½ miles north of the town of Jackson, Clarke County, Alabama, U.S.A.

Diagnosis Test low trochospiral, spiral side tending to be slightly convex, umbilical side convex; equatorial periphery moderately lobulate, axial periphery broadly rounded.
Wall perforate, surface pitted in early chambers to finely pitted in last chamber.
Chambers subspherical, somewhat compressed laterally, arranged in about three whorls; the four chambers of the last whorl increase fairly rapidly in size.
Sutures on spiral side slightly curved, depressed; on umbilical side radial, depressed.
Umbilicus fairly narrow.
Aperture a high, distinct arch, bordered by a rim, interiomarginal, extraumbilical-umbilical, extending almost to the periphery.

Strat. distr. Upper part of *Truncorotaloides rohri* zone into upper part of *Globigerina ampliapertura* zone.

Remarks See under *Globigerina ampliapertura* BOLLI.
Locality of figured specimen is the above-mentioned type locality.

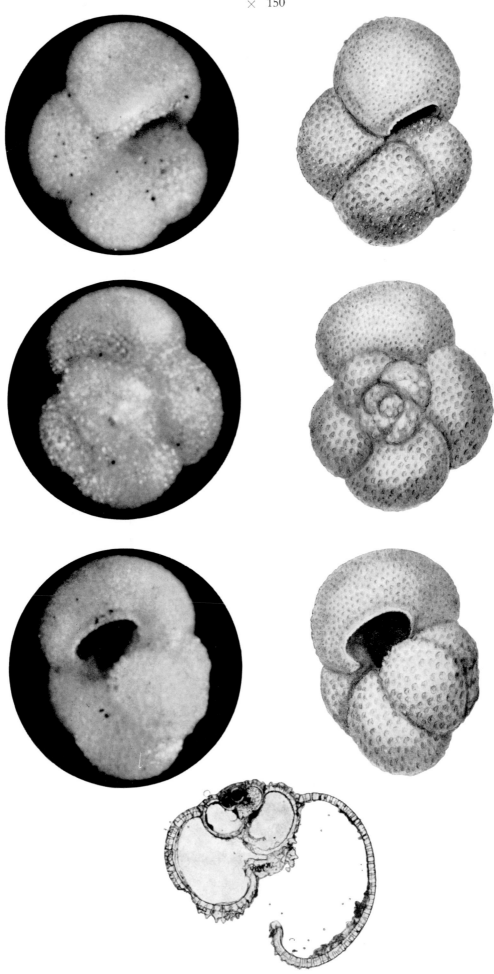

Globorotalia laevigata BOLLI

Reference *Globorotalia pusilla laevigata* BOLLI, 1957: The Genera *Globigerina* and *Globorotalia* in the Paleocene-Lower Eocene Lizard Springs Formation of Trinidad, B.W.I. — United States National Museum Bulletin, 215:78, pl. 20, figs. 5-7.

Type locality North-east bank of Tank Farm at the old Club Site, Pointe-à-Pierre, Trinidad.

Diagnosis Test low trochospiral, biconvex, compressed; equatorial periphery nearly circular, slightly lobulate, axial periphery acute, often with a faint keel.
Wall finely perforate, surface of early chambers slightly nodose, that of last chambers smooth.
Chambers strongly compressed, arranged in about 3 whorls; the six to seven chambers of the last whorl increase slowly in size.
Sutures on spiral side strongly curved, limbate, on umbilical side radial, slightly depressed to flush.
Umbilicus narrow.
Aperture a low, interiomarginal, extraumbilical-umbilical arch, bordered by a lip.

Strat. distr. Ranging throughout the *Globorotalia pseudomenardii* zone.

Remarks *Globorotalia laevigata* BOLLI may be synonymous with *Globorotalia albeari* CUSHMAN and BERMUDEZ. According to the original description, this latter species should have nine to ten chambers in the last whorl. One of the type-figures, however, suggests a smaller number.
Locality of figured specimen is well San Roman I, Guatemala, depth 1870—1880 feet.

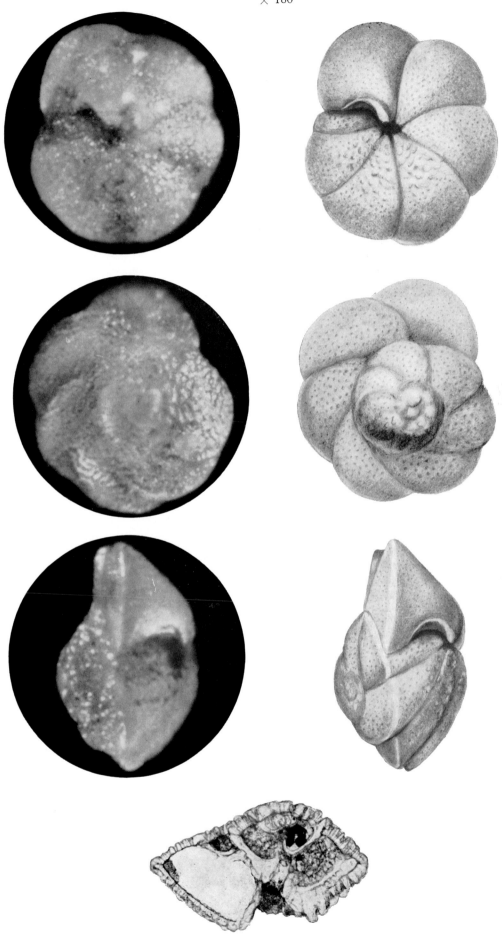

Globorotalia lehneri CUSHMAN and JARVIS

Reference   *Globorotalia lehneri* CUSHMAN and JARVIS, 1929: New Foraminifera from Trinidad. —
Contributions from the Cushman Laboratory for Foraminiferal Research, 5 (1):17, pl.3,
fig. 16a-c.

Type locality   Near source of Morugo River, Trinidad.

Diagnosis   Test low trochospiral, biconvex, compressed; equatorial periphery strongly lobulate,
axial periphery acute with a spinose or rugose keel, which protrudes at the middle of
the periphery of each chamber.
Wall finely perforate, surface of early chambers spinose to rugose, that of later chambers
smooth.
Chambers angular, slightly inflated, compressed, arranged in 2½-3 whorls; the six to
seven chambers of the last whorl increase moderately in size.
Sutures on spiral side slightly curved between early chambers to radial between last
chambers, depressed, on umbilical side radial, depressed.
Umbilicus narrow, shallow.
Aperture an interiomarginal, extraumbilical-umbilical arched slit.

Strat. distr.   Upper part of *Globigerapsis kugleri* zone to top *Truncorotaloides rohri* zone.

Remarks   Locality of figured specimen is Dyr el Kef, Tunisia.

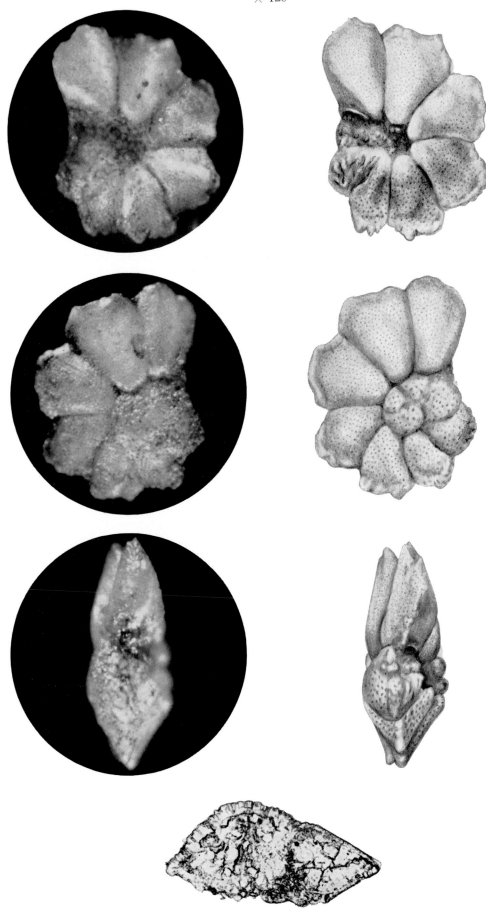

Globorotalia mckannai (WHITE)

Reference   *Globigerina mckannai* WHITE, 1928: Some index Foraminifera of the Tampico embayment area of México. — Journal of Paleontology, 2 (3):194, pl.27, fig. 16a-c.

Type locality   Columbus station of the Tampico Monterey railroad line, Tampico embayment, Mexico.

Diagnosis   Test low trochospiral, spiral side slightly convex, umbilical side strongly convex, inflated; equatorial periphery nearly circular, axial periphery broadly rounded.
Wall finely perforate, surface of most chambers spinose or nodose, surface of last chambers less so or smooth.
Chambers inflated, fairly compressed laterally, arranged in 2-3 whorls; the chambers of the last whorl, usually six in number, increase moderately in size.
Sutures on spiral side slightly curved to oblique, depressed; on umbilical side radial, depressed.
Umbilicus narrow, deep.
Aperture a low interiomarginal, extraumbilical-umbilical arch.

Strat. distr.   Ranging throughout upper part of the *Globorotalia pseudomenardii* zone, extending a little into the lower part of this zone.

Remarks   Locality of figured specimen is GS 1574, Velasco Shale, Mexico.

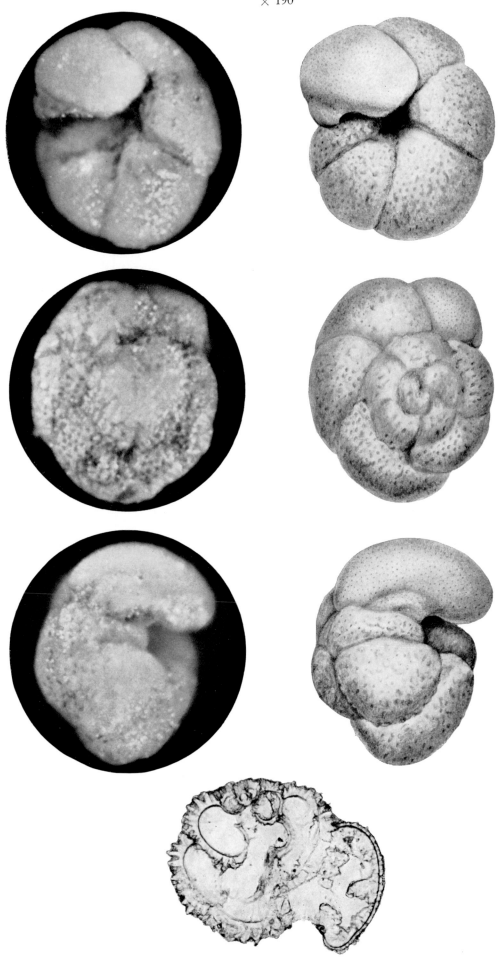

Globorotalia pseudobulloides (Plummer)

Reference	*Globigerina pseudo-bulloides* Plummer, 1926: Foraminifera of the Midway Formation in Texas. — University of Texas Bulletin, 2644:133, pl.VIII, fig. 9a-c.
Type locality	Station 23, shallow ditch at road corner south-east of new Corsicana reservoir on the road to Mildred, Navarro County, Texas, U.S.A.
Diagnosis	Test very low trochospiral, spiral side flattened, initial whorl either slightly depressed or somewhat convex, umbilical side convex; equatorial periphery distinctly lobulate, axial periphery rounded. Wall finely perforate, surface smooth. Chambers inflated, globular to subglobular, arranged in $2\frac{1}{2}$ whorls; the five chambers of the last whorl increase fairly rapidly in size. Sutures on spiral side curved, less so in the last chambers, depressed; on umbilical side radial, depressed. Umbilicus fairly narrow, open. Aperture a rather low, interiomarginal, extraumbilical-umbilical arch, bordered by a distinct lip; aperture of penultimate chamber occasionally visible.
Strat. distr.	Base of *Globigerina daubjergensis* zone to top of *Globorotalia angulata* zone.
Remarks	*Globorotalia varianta* (Subbotina) appears to be a closely related species. Locality of figured specimen is the above-mentioned type locality.

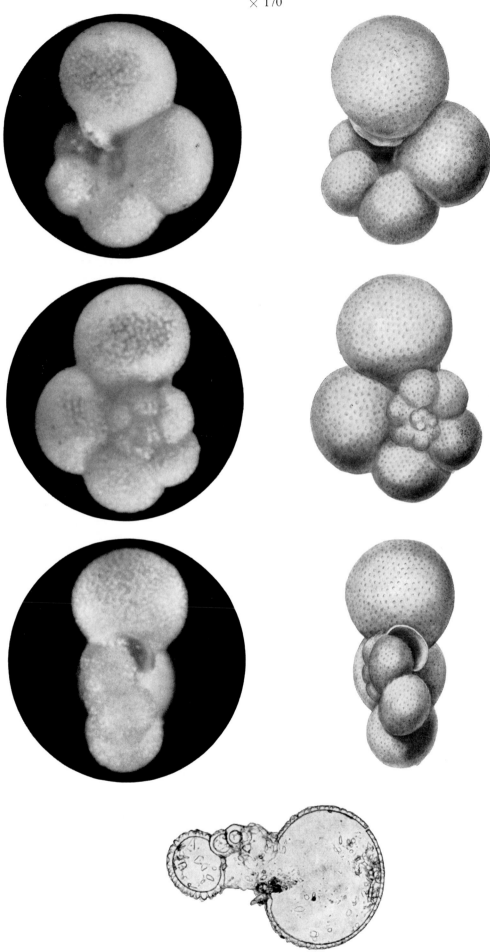

Globorotalia pseudomenardii BOLLI

Reference   *Globorotalia pseudomenardii* BOLLI, 1957: The Genera *Globigerina* and *Globorotalia* in the Paleocene-Lower Eocene Lizard Springs Formation of Trinidad, B.W.I. — United States National Museum Bulletin, 215:77, pl.20, figs. 14-17.

Type locality   Trinidad Leaseholds, Ltd., well Guayaguayare 159, Trinidad, depth 4324-4330 feet.

Diagnosis   Test very low trochospiral, biconvex; equatorial periphery slightly lobulate, axial periphery acute with a distinct keel.
Wall finely perforate, surface smooth.
Chambers strongly compressed, arranged in about three whorls; the chambers of the last whorl, commonly five in number, increase fairly rapidly in size.
Sutures on spiral side strongly curved, especially so between last chambers, slightly depressed to flush, on umbilical side radial, depressed.
Umbilicus fairly narrow, shallow.
Aperture interiomarginal, extraumbilical-umbilical, a low arch, bordered by a distinct lip.

Strat. distr.   Ranging throughout the *Globorotalia pseudomenardii* zone.

Remarks   For many years this species has been determined and designated in publications as *Globorotalia membranacea* (EHRENBERG). The latter name, however, is invalid because EHRENBERG (1854) figured under *Planulina membranacea* the spiral views of two rotalid Foraminifera from the Cretaceous that are at least specifically different. Furthermore, no description or depository of a holotype was given by the author.
Locality of figured specimen is DB 115, Trinidad.

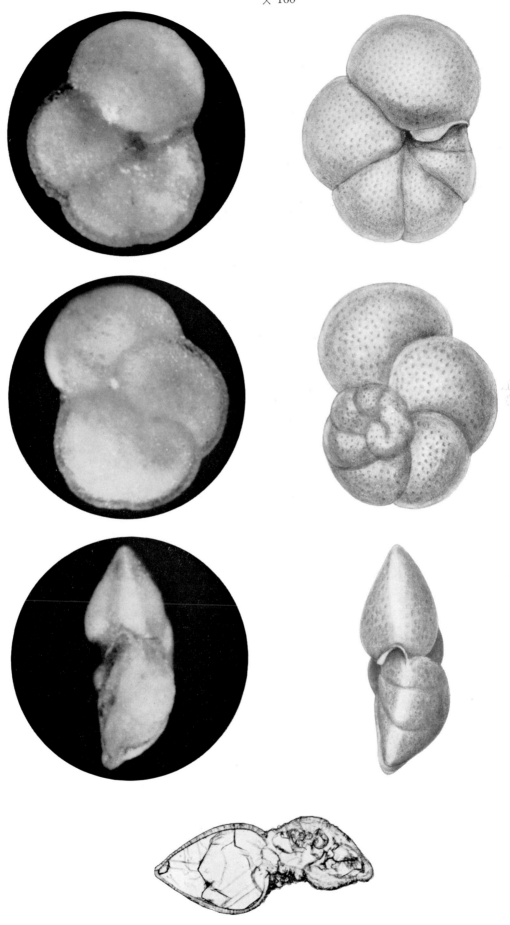

Globorotalia pusilla BOLLI

Reference *Globorotalia pusilla pusilla* BOLLI, 1957: The Genera *Globigerina* and *Globorotalia* in the Paleocene-Lower Eocene Lizard Springs Formation of Trinidad, B.W.I. — United States National Museum Bulletin, 215:78, pl.20, figs. 8-10.

Type locality Trinidad Leaseholds, Ltd., well Guayaguayare 159, Trinidad, depth 4778-4790 feet.

Diagnosis Test low trochospiral, biconvex, compressed; equatorial periphery lobulate, axial periphery acute to subacute.
Wall finely perforate, surface smooth except sometimes in the early chambers which may be very slightly nodose.
Chambers compressed, arranged in 2½-3 whorls; the five chambers of the last whorl increase moderately in size.
Sutures on spiral side strongly curved, slightly depressed, on umbilical side radial, depressed.
Umbilicus narrow.
Aperture a low, interiomarginal, extraumbilical-umbilical arch, bordered by a narrow lip.

Strat. distr. Base of *Globorotalia angulata* zone into lower part of *Globorotalia pseudomenardii* zone.

Remarks This species differs from the closely related *Globorotalia laevigata* BOLLI in having fewer chambers in the last whorl, more depressed sutures and in being slightly more lobulate. Locality of figured specimen is Dyr el Kef, W. Tunisia.

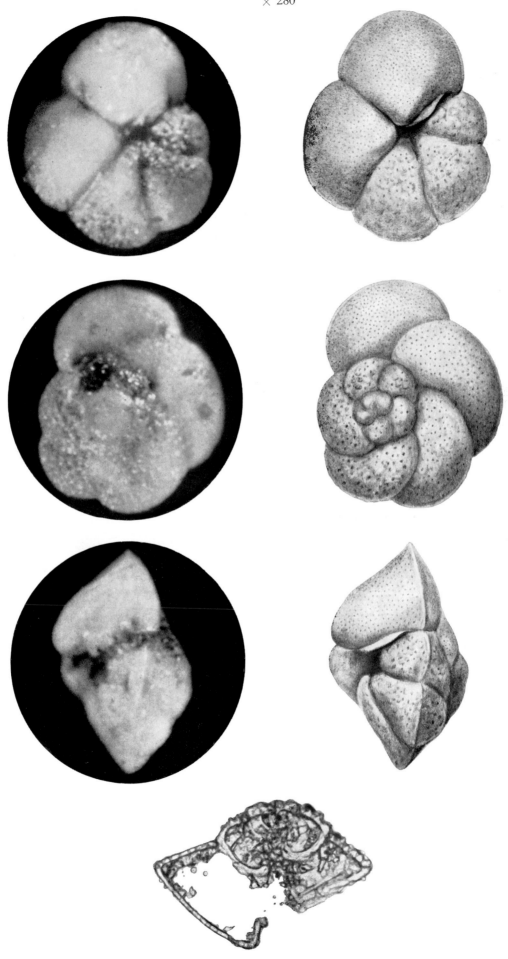

Globorotalia renzi BOLLI

Reference   *Globorotalia renzi* BOLLI, 1957: Planktonic Foraminifera from the Eocene Navet and San Fernando Formations of Trinidad, B.W.I. — United States National Museum Bulletin, 215:168, pl.38, fig. 3a-c.

Type locality   Holotype from the Middle Eocene, Navet Formation; a block in the Upper Oligocene-Lower Miocene Nariva Formation, in cutting west of tank 127, north of The Avenue and 850 feet west of its junction with Bon Accord Road, Pointe-à-Pierre, Trinidad.

Diagnosis   Test small, very low trochospiral, somewhat biconvex; equatorial periphery almost circular, axial periphery angular with a thin keel.
Wall finely perforate, surface of early chambers slightly spinose or nodose, surface of last chambers smooth or finely pitted.
Chambers strongly compressed, arranged in about 2½ whorls; the six to seven chambers of the last whorl increase gradually in size.
Sutures on spiral side curved, slightly depressed to somewhat raised between last chambers; on umbilical side very slightly curved to radial, slightly depressed between last chambers.
Umbilicus very narrow, shallow.
Aperture a low, interiomarginal, extraumbilical-umbilical arch, bordered by a lip.

Strat. distr.   Base of *Globorotalia bullbrooki* zone to top of *Truncorotaloides rohri* zone.

Remarks   Locality of figured specimen is the above-mentioned type locality.

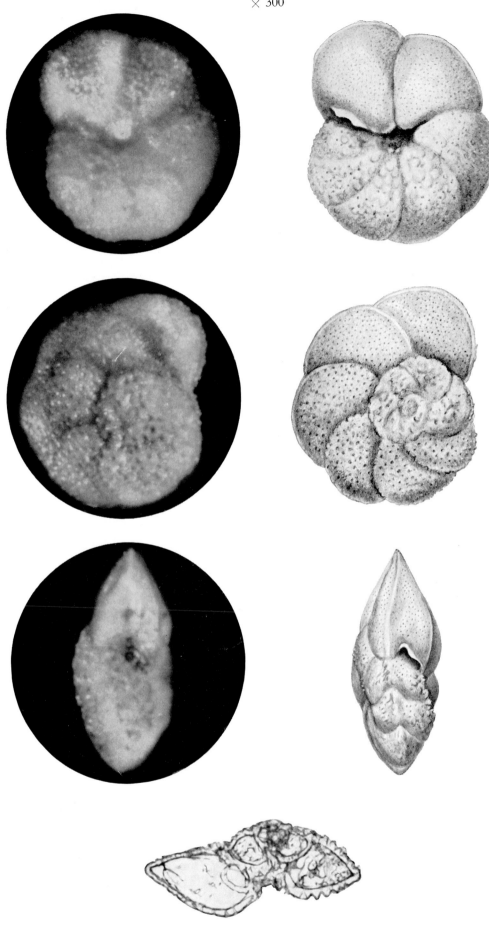

Globorotalia rex MARTIN

Reference   *Globorotalia rex* MARTIN, 1943: Eocene Foraminifera from the type Lodo Formation, Fresno County, California. — Stanford University Publications, University Series, Geological Sciences, III (3):117, pl.VIII, fig. 2a-c.

Type locality   Stanford University Locality No. M-74, type section of the Lodo Formation, Fresno County, California, U.S.A.

Diagnosis   Test very low trochospiral, spiral side flat or slightly convex, umbilical side markedly convex; equatorial periphery lobulate, axial periphery angular with distinct peripheral keel, often ornamented with short spines or knobs.
Wall perforate, surface coarsely spinose or nodose, surface of last chamber less so.
Chambers angular, inflated, arranged in 2-2½ whorls; the chambers of the last whorl, usually four in number, increase rapidly in size, the last chamber constituting a considerable part of the surface of the test.
Sutures on spiral side strongly curved, depressed; on umbilical side radial, distinctly depressed.
Umbilicus narrow, deep.
Aperture a low, interiomarginal, extraumbilical-umbilical arch, bordered by a faint lip.

Strat. distr.   Ranging throughout the *Globorotalia rex* zone and the lower part of the *Globorotalia formosa/aragonensis* zone.

Remarks   *Globorotalia rex* MARTIN differs from the related *Globorotalia aequa* CUSHMAN and RENZ in being more robust and in having a thickened peripheral keel.
Locality of figured specimen is HK 1831, Trinidad.

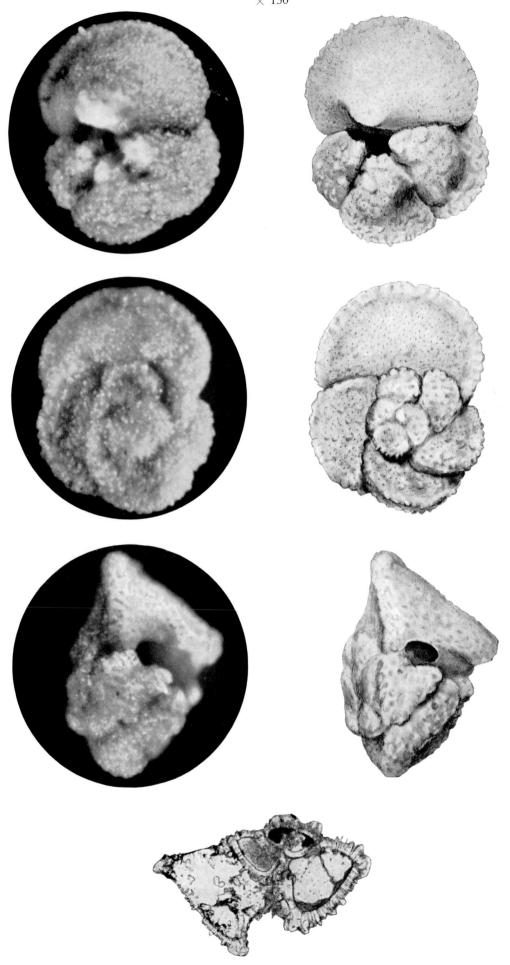

Globorotalia spinulosa CUSHMAN

Reference *Globorotalia spinulosa* CUSHMAN, 1927: New and interesting Foraminifera from Mexico and Texas. — Contributions from the Cushman Laboratory for Foraminiferal Research, 3 (2):114, pl.23, fig. 4a-c.

Type locality Alazan clay; Rio Tuxpan, crossing of road from Palo Blanco to La Noria and along Rio Pantepec about 200 metres above its mouth, Vera Cruz, Mexico.

Diagnosis Test very low trochospiral, spiral side almost flat to slightly convex, umbilical side convex; equatorial periphery lobulate, axial periphery acute with a prominent keel ornamented with spines or knobs.
Wall finely perforate, surface of early chambers on spiral side nodose, later ones smooth, surface of early chambers on spiral side nodose to rugose at the umbilical edge, last chamber smooth.
Chambers angular, inflated, strongly compressed at periphery, arranged in 2½ whorls; the four to five chambers of the last whorl increase rapidly in size.
Sutures on spiral side curved, depressed; on umbilical side radial, depressed.
Umbilicus fairly narrow to fairly wide.
Aperture a medium, interiomarginal, extraumbilical-umbilical arch, bordered by a lip.

Strat. distr. Base of *Globorotalia bullbrooki* zone to top of *Orbulinoides beckmanni* zone.

Remarks Locality of figured specimen is BG 8581, Trinidad.

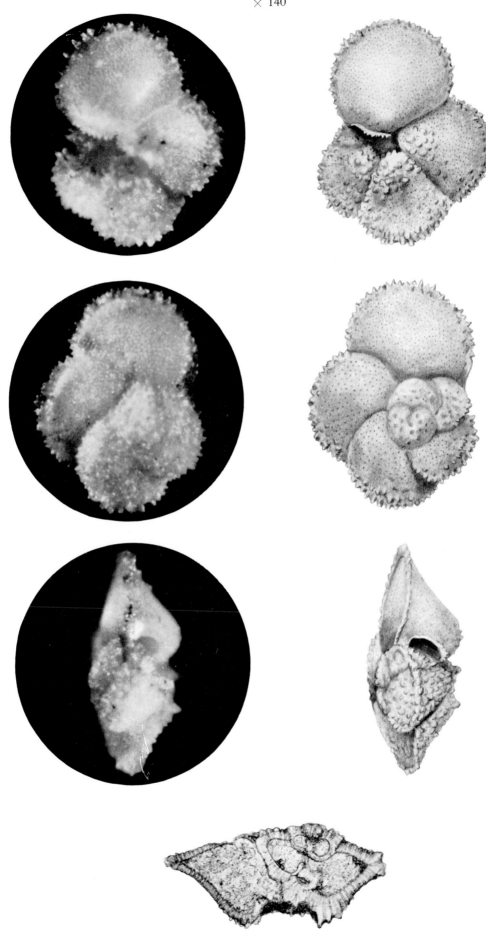

Globorotalia trinidadensis BOLLI

Reference *Globorotalia trinidadensis* BOLLI, 1957: The Genera *Globigerina* and *Globorotalia* in the Paleocene-Lower Eocene Lizard Springs Formation of Trinidad, B.W.I. — United States National Museum Bulletin, 215:73, pl.16, figs. 19-23.

Type locality Trinidad Petroleum Development well Moruga 3, Trinidad, depth 10, 259-10, 261 feet.

Diagnosis Test very low trochospiral, inflated; equatorial periphery lobulate, axial periphery rounded.
Wall finely perforate, surface of early chambers slightly rugose, that of later chambers smooth.
Chambers inflated, globular to subglobular, arranged in about $2\frac{1}{2}$ whorls; the five to seven chambers of the last whorl increase slowly in size.
Sutures on spiral side slightly curved to radial, depressed; on umbilical side radial, depressed.
Umbilicus fairly wide, open.
Aperture a rather low, interiomarginal, extraumbilical-umbilical arch, bordered by a lip-like flap, part of which can be seen in the penultimate and preceding chambers.

Strat. distr. Ranging throughout the *Globigerina daubjergensis* zone and the lower part of the *Globorotalia uncinata* zone.

Remarks *Globorotalia trinidadensis* differs from *Globorotalia pseudobulloides* (PLUMMER) in its larger size and in having more chambers in the final whorl. Early chambers may show a rugose surface.
Locality of figured specimen is Dyr el Kef, Tunisia.

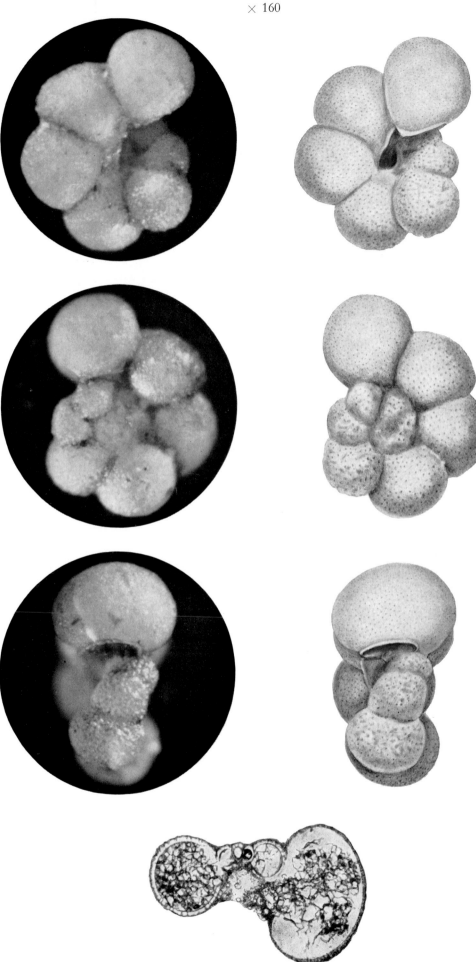

216

Globorotalia uncinata BOLLI

Reference
Globorotalia uncinata BOLLI, 1957: The Genera *Globigerina* and *Globorotalia* in the Paleocene-Lower Eocene Lizard Springs Formation of Trinidad, B.W.I. — United States National Museum Bulletin, 215:74, pl.17, figs. 13-15.

Type locality
West side of railway track south of the Point-à-Pierre railway station, about 500 feet from the level crossing of Station Road, Point-à-Pierre, Trinidad.

Diagnosis
Test low trochospiral, spiral side almost flat or slightly convex; equatorial periphery distinctly lobulate, axial periphery rounded to subangular.
Wall perforate, surface partly spinose to slightly rugose.
Chambers subangular, inflated, laterally compressed, arranged in 2½-3 whorls; the five to six chambers of the last whorl increase moderately in size.
Sutures on spiral side strongly curved, depressed; on umbilical side radial, depressed.
Umbilicus fairly narrow, deep.
Aperture an interiomarginal, extraumbilical-umbilical arch, bordered by a lip.

Strat. distr.
Ranging throughout the *Globorotalia uncinata* zone and the lower part of the *Globorotalia angulata* zone.

Remarks
Globorotalia uncinata BOLLI differs from the related *Globorotalia pseudobulloides* (PLUMMER) in having subangular, laterally distinctly truncated chambers and more strongly curved sutures on the spiral side. *Globorotalia uncinata* BOLLI is regarded as the ancestor of *Globorotalia angulata* (WHITE).
Locality of figured specimen is the above-mentioned type locality.

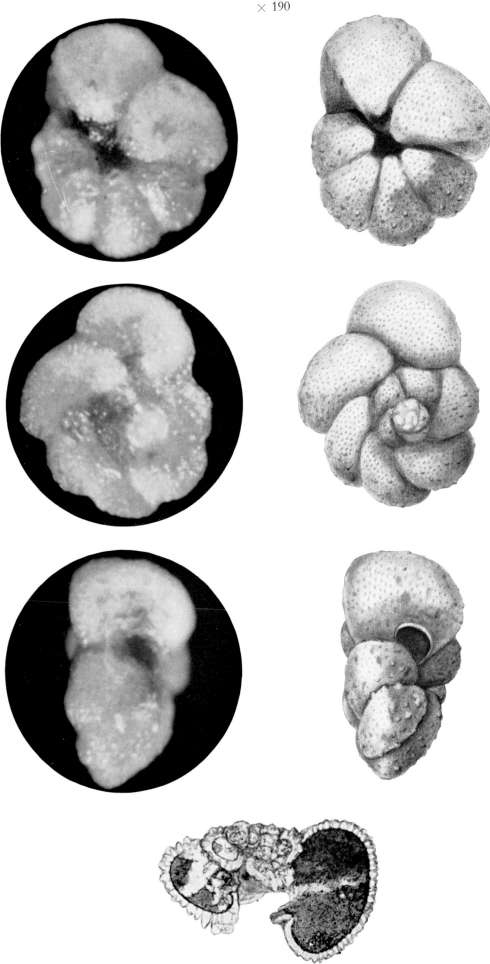

Globorotalia velascoensis (CUSHMAN)

Reference *Pulvinulina velascoensis* CUSHMAN, 1925: Some new Foraminifera from the Velasco Shale of Mexico. — Contributions from the Cushman Laboratory for Foraminiferal Research, 1 (1):19, pl.3, fig. 5a-c.

Type locality Holotype (Cushman Coll. No. 4347) from the Velasco Shale, Tamalte Arroyo, Hacienda El Limon, State of San Luis Potosi, Mexico.

Diagnosis Test very low trochospiral, spiral side almost flat, umbilical side strongly convex; the outer wall of the chambers of the last whorl is usually somewhat concave; equatorial periphery slightly lobulate to nearly circular, axial periphery angular with distinct peripheral keel, often ornamented with short spines or knobs.
Wall finely perforate, surface smooth, around umbilical area rugose.
Chambers angular, inflated, arranged in 2½-3 whorls, the five to seven chambers of the last whorl increase moderately or feebly in size.
Sutures on spiral side curved, somewhat raised and slightly beaded; on umbilical side radial, depressed.
Umbilicus wide, deep.
Aperture a low, interiomarginal, extraumbilical-umbilical arch, bordered by a faint lip; apertures of penultimate and preceding chambers visible in well preserved specimens.

Strat. distr. Upper part of *Globorotalia angulata* zone to top of *Globorotalia velascoensis* zone.

Remarks The size of the species may vary considerably; smaller specimens have been assigned to *Globorotalia wilcoxensis* var. *acuta* TOULMIN, which is a synonym of *Globorotalia velascoensis* (CUSHMAN).
Locality of figured specimen is the above-mentioned type locality.

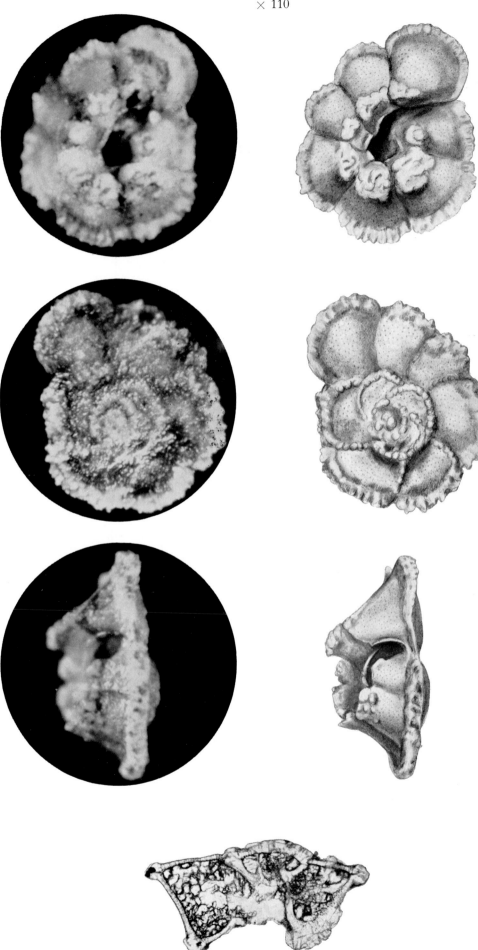

Globorotalia wilcoxensis CUSHMAN and PONTON

Reference   *Globorotalia wilcoxensis* CUSHMAN and PONTON, 1932: An Eocene foraminiferal fauna of Wilcox Age from Alabama. — Contributions from the Cushman Laboratory for Foraminiferal Research, 8 (3):71, pl.9, fig. 10a-c.

Type locality   Railroad cut one mile N. of Ozark, Alabama, U.S.A.

Diagnosis   Test very low trochospiral, spiral side flat, umbilical side convex, inflated; equatorial periphery lobulate, axial periphery rounded, in last chamber subacute.
Wall perforate, surface distinctly nodose, that of last chamber less so.
Chambers inflated, slightly compressed laterally, arranged in about $2\frac{1}{2}$ whorls; the four chambers of the last whorl increase rapidly in size.
Sutures on spiral side slightly curved, depressed; on umbilical side radial, depressed. Umbilicus fairly narrow, deep.
Aperture a semicircular interiomarginal, extraumbilical-umbilical arch, bordered by a narrow lip.

Strat. distr.   Ranging throughout the *Globorotalia rex* zone.

Remarks   Locality of figured specimen is the above-mentioned type locality.

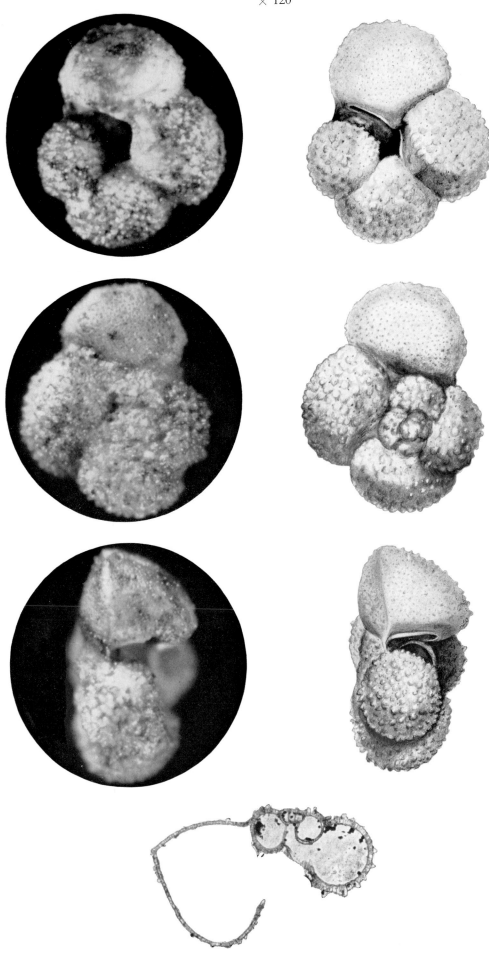

Hantkenina aragonensis NUTTALL

Reference *Hantkenina mexicana* CUSHMAN var. *aragonensis* NUTTALL, 1930: Eocene Foraminifera from Mexico. — Journal of Paleontology, 4 (3):284, pl.24, figs. 1-3.

Type locality Aragon Formation near La Antigua, State of Vera Cruz, Mexico.

Diagnosis Test planispiral, involute, biumbilicate.
Wall perforate, surface pitted.
Chambers strongly radial elongate, terminating in a spine, compressed; the chambers of the last whorl, usually five in number, are stellate in arrangement, distinctly separated, increasing rapidly in size; the spines are situated in the prolongation of the chamber axis.
Sutures straight or only slightly curved, depressed.
Aperture interiomarginal, equatorial, triradiate, two of the rays forming a slit across the base of the final chamber face, the third ray originating from the centre of this slit and extending up, the rays bordered by an apertural flange.

Strat. distr. Ranging throughout the *Globorotalia bullbrooki* zone and the lower part of the *Globigerapsis kugleri* zone.

Hantkenina mexicana CUSHMAN

Reference *Hantkenina mexicana* CUSHMAN, 1924: A new genus of Eocene Foraminifera. — Proceedings United States National Museum, 66 (2567), art. 30:3, pl.2, fig. 2.

Type locality La Laja, Zardo Creek, one kilometre south-west of Tierra, Colorado, Mexico.

Diagnosis Test planispiral, involute, biumbilicate.
Wall perforate, surface pitted.
Chambers radial elongate, terminating in a spine, strongly compressed; usually six chambers in the last whorl, the first of which are in a stellate arrangement and the latter ones disposed tangentially, the spines, being situated at the anterior angle of the chambers.
Sutures straight or only slightly curved, depressed.
Aperture as in *Hantkenina aragonensis*.

Strat. distr. Ranging throughout the *Globigerapsis kugleri* zone and the upper part of the *Globorotalia bullbrooki* zone, and extending a little into the lower part of the latter one.

Hantkenina dumblei WEINZIERL and APPLIN

Reference *Hantkenina dumblei* WEINZIERL and APPLIN, 1929: The Claiborne Formation on the Coastal Domes. — Journal of Paleontology, 3:402, pl.3, fig. 5a-b.

Type locality Rio Bravo Oil Co., Deussen B1, 4010 feet, South Liberty, Liberty County, Texas, U.S.A.

Diagnosis Test planispiral, involute, biumbilicate.
Wall perforate, surface smooth.
Chambers elongate, terminating in a short or stout spine, compressed; the five to six chambers of the last whorl are tangentially arranged, the spines situated at the anterior angle of the chambers.
Sutures straight or slightly sigmoidally curved, depressed.
Aperture as in *Hantkenina aragonensis*.

Strat. distr. Ranging throughout the *Globigerapsis kugleri* zone and the lower part of the *Globorotalia lehneri* zone.

Remarks The three *Hantkenina* species described above belong to the group with chambers longer than broad. *Hantkenina mexicana* CUSHMAN differs from *Hantkenina aragonensis* NUTTALL in that the chambers tend to become tangentially arranged, with more chambers in the last whorl becoming somewhat less distinctly separated.
Hantkenina dumblei WEINZIERL and APPLIN, however, differs from *Hantkenina mexicana* mainly in having all chambers of the last whorl tangentially arranged and even less distinctly separated.
The figured specimens of *Hantkenina aragonensis* and *Hantkenina mexicana* are from different localities in the Caribbean area. The figured specimens of *Hantkenina dumblei* are from F 826, Tunisia.

Hantkenina alabamensis CUSHMAN

Reference   *Hantkenina alabamensis* CUSHMAN, 1924: A new genus of Eocene Foraminifera. — Proceedings United States National Museum, 60 (30) :1, pl.1, figs. 1-6.

Type locality   Zeuglodon bed at Cocoa post office, Alabama, U.S.A.

Diagnosis   Test planispiral, biumbilicate, almost involute; equatorial periphery slightly lobulate, axial periphery rounded.

Wall finely perforate, surface of early chambers may be finely hispid, surface of later chambers smooth.

Chambers moderately inflated, early ones subglobular, later ones more or less ovate, with a long peripheral tubulospine at the forward margin of each chamber; the last whorl consists of five to six chambers; the last chambers of the final whorl are attached near the base of the spines and may partially envelop the spine of the preceding chamber. Sutures radial, depressed.

Aperture interiomarginal, equatorial, triradiate, two of the rays forming a slit across the base of the final chamber face, the third originating from the centre of this slit and extending up the face towards the peripheral spine, flaring slightly to become rounded at its upper end, the rays bordered by apertural flanges.

Strat. distr.   Upper part of *Truncorotaloides rohri* zone to top of *Globigerapsis mexicana* zone.

Remarks   This species differs from *Hantkenina mexicana* CUSHMAN in having more inflated chambers. The spines of the latter species are never enveloped by a chamber wall.

Locality of figured specimen is the above-mentioned type locality.

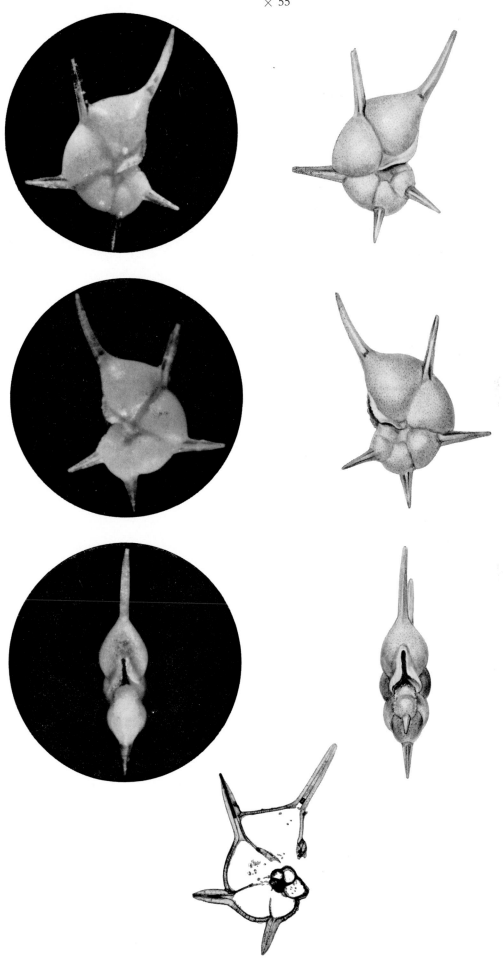

Hantkenina brevispina CUSHMAN

Reference   *Hantkenina brevispina* CUSHMAN, 1924: A new genus of Eocene Foraminifera. — Proceedings United States National Museum, 60 (30):1, pl.2, fig. 3.

Type locality   Rio Pantepec, 2.2 km S.20° W. from Buena Vista, Mexico.

Diagnosis   Test planispiral, biumbilicate, almost involute; equatorial periphery slightly lobulate to nearly circular, axial periphery rounded.
Wall finely perforate, surface smooth.
Chambers inflated, subglobular, with a short tubulospine at the forward margin of each chamber; the six succeeding chambers of the last whorl are attached near the base of the spines and may partially or completely envelop the spine of the preceding chamber.
Sutures almost radial, depressed.
Aperture interiomarginal, equatorial, triradiate, two of the rays forming an arch across the base of the final chamber face, the third ray originating from the centre of this arch and extending up the face towards the peripheral spine, flaring slightly to become rounded at its upper end, the rays bordered by apertural flanges.

Strat. distr.   Ranging throughout the *Globigerapsis mexicana* zone and the lower part of the *Globorotalia cerroazulensis* zone.

Remarks   This species is probably the ancestor of *Cribrohantkenina bermudezi*. It differs in having more chambers in the last whorl and in having the typical triradiate aperture.
Locality of figured specimen is near Cerro Azul, Mexico.

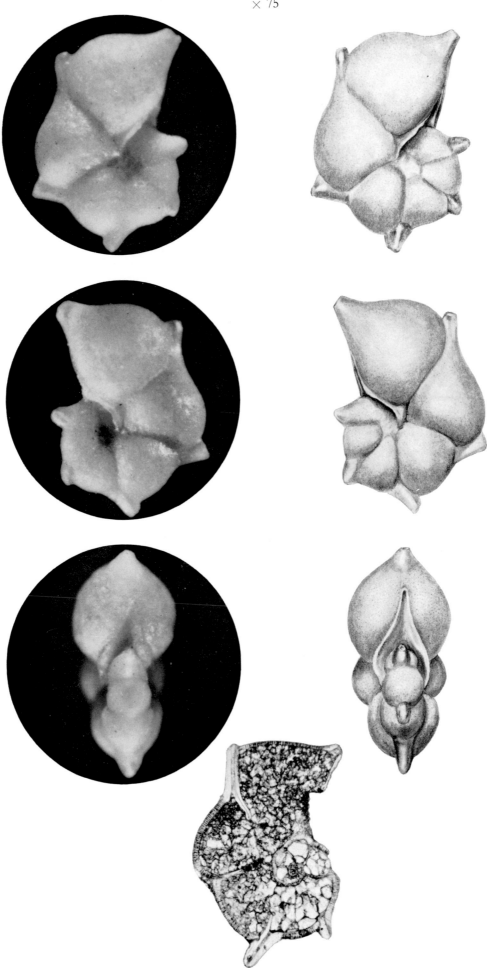

Hastigerina micra (Cole)

Reference *Nonion micrus* Cole, 1927: A foraminiferal fauna from the Guayabal Formation in Mexico. — Bulletins of American Paleontology, XIV (51):22, pl.5, fig. 12.

Type locality Guayabal Formation-Cliff outcropping twelve kilometres due east of the village of Potrero in the state of Vera Cruz, Mexico.

Diagnosis Test planispiral, early stage may be slightly trochospiral; biumbilicate; equatorial periphery slightly lobulate to lobulate, axial periphery rounded.
Wall finely perforate, surface smooth.
Chambers inflated, globular to subglobular in the early stage, ovate in the later stage; six to eight chambers in the last whorl, the last chambers of the final whorl increase sometimes very rapidly in size.
Sutures curved, depressed.
Aperture a broad, interiomarginal, equatorial arch, bordered by a lip.

Strat. distr. Base of *Globorotalia formosa/aragonensis* zone to top of *Globorotalia cerroazulensis* zone. Questionable occurrences in upper part of *Globorotalia rex* zone and in lower part of *Globigerina ampliapertura* zone.

Remarks Glaessner (1937) changed the generic status of this species to *Globigerinella* which is now regarded as a junior synonym of *Hastigerina*.
Locality of figured specimen is the above-mentioned type locality.

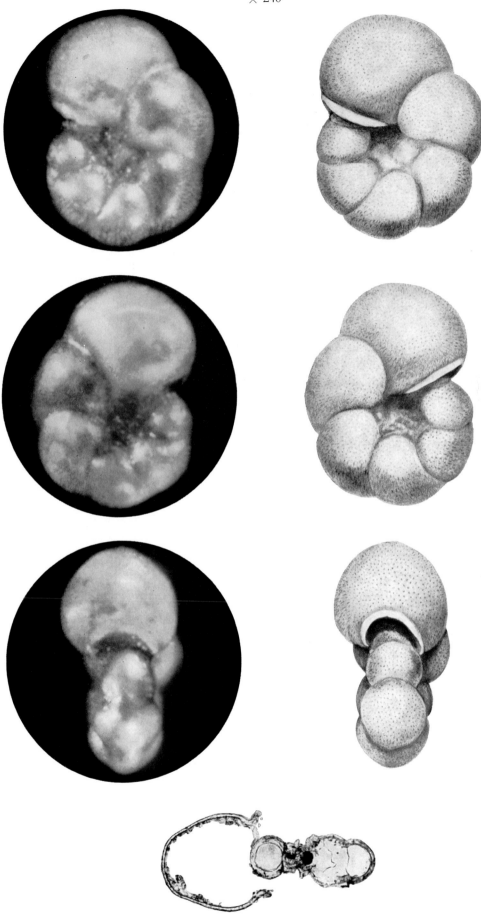

Orbulinoides beckmanni (SAITO)

Reference   *Porticulasphaera beckmanni* SAITO, 1962: Eocene planktonic foraminifera from Hahajima (Hillsborough Island). – Pal. Soc. Japan, Trans. Proc., n. ser., 45: 221, pl. 34, figs. 1a-2.

Type locality   Hahajima, Hillsborough Island, Bonin Group, Japan.

Diagnosis   Test subglobular, early portion low trochospiral; periphery practically circular.
Wall coarsely perforate, thick, surface smooth, slightly pitted.
Chambers subglobular, compressed laterally, arranged in about three whorls, about five chambers per whorl, the final chamber much inflated and strongly embracing and enveloping the umbilical region of the early coil, larger in size than the entire previous portion of the test.
Sutures almost radial, depressed.
Primary aperture in the early portion interiomarginal, umbilical, with small secondary sutural openings on the spiral side as in *Globigerinoides,* the primary aperture of the early portion covered by the final embracing chamber, which has only the numerous small sutural secondary apertures completely encircling its basal margin.

Strat. distr.   Ranging throughout the *Orbulinoides beckmanni* zone.

Remarks   This species has been confused with *Globigerapsis mexicana* (CUSHMAN). See SAITO in Pal. Soc. Japan, Trans. Proc., n. ser., 45: 209–225, 1962 and BLOW and SAITO in Micropaleontology, 14 (3): 357-360, 1968.
Locality of figured specimen is DB 276, Trinidad.

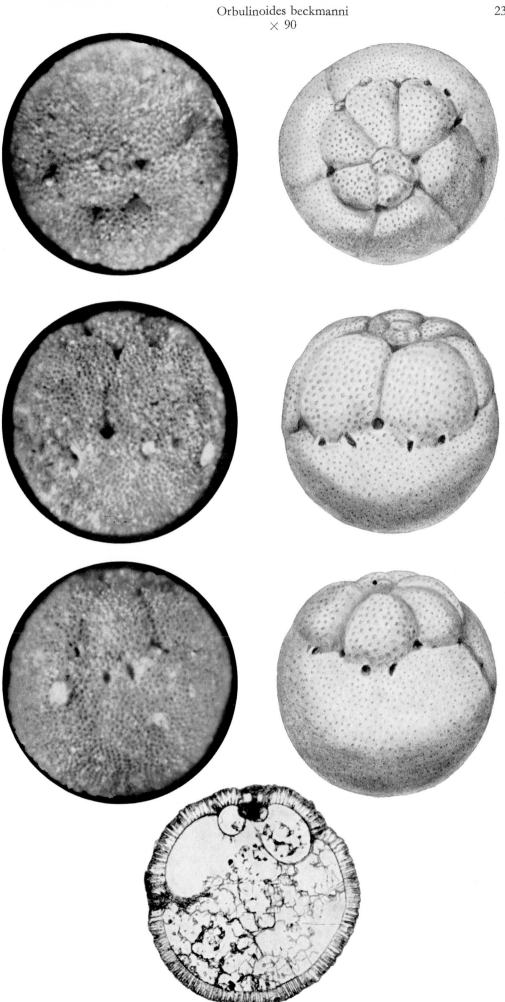

Truncorotaloides rohri Brönnimann and Bermudez

Reference	*Truncorotaloides rohri* Brönnimann and Bermudez 1953: *Truncorotaloides,* a new foraminiferal genus from the Eocene of Trinidad, B.W.I. — Journal of Paleontology, 27 (6):818, pl.87, figs. 7-9.
Type locality	A marl pebble bed situated in the Duff road area, near Kelly junction, about 7 miles east of Pointe-à-Pierre, central Trinidad.
Diagnosis	Test low trochospiral, spiral side flat to slightly convex, umbilical side convex; equatorial periphery lobulate, axial periphery rounded to subangular. Wall perforate, surface prominently spinose throughout. Chambers subglobular to subangular, inflated, compressed laterally, arranged in about three whorls; the five chambers of the last whorl increase rapidly in size, somewhat overlapping on the spiral side. Sutures on spiral side curved to radial between last chambers, depressed. Umbilicus fairly narrow, deep. Primary aperture a medium, interiomarginal, extraumbilical-umbilical arch, bordered by a rim, with about two single secondary sutural apertures visible on the spiral side at the inner margin of the later chambers where they lie against the previous whorl.
Strat. distr.	Base of *Globigerapsis kugleri* zone to top of *Truncorotaloides rohri* zone. Questionable occurrence in upper part of *Globorotalia bullbrooki* zone.
Remarks	In addition to *Truncorotaloides rohri,* Brönnimann and Bermudez (1953) described three varieties of this species which illustrate the variation of chamber and test shape ranging from rounded to angular forms. Locality of figured specimen is Hg 858, Trinidad.

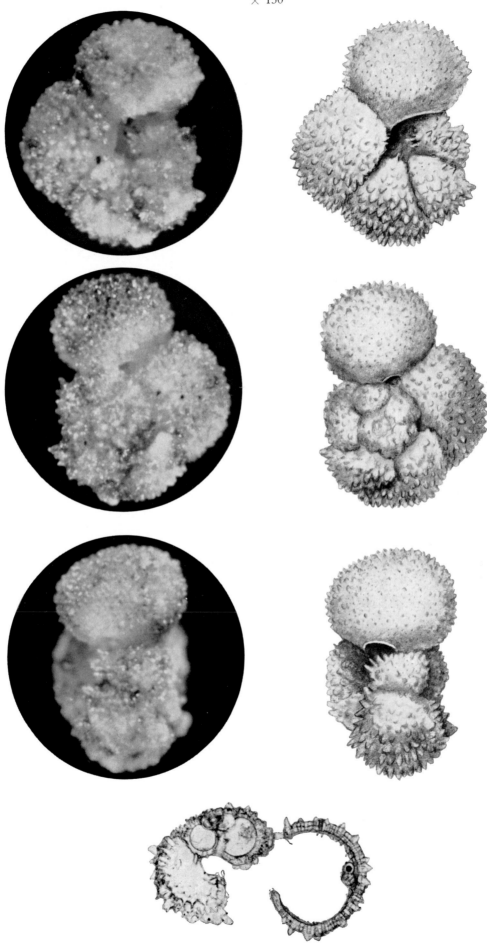

Truncorotaloides topilensis (Cushman)

Reference *Globigerina topilensis* Cushman, 1925: New Foraminifera from the Upper Eocene of Mexico. — Contributions from the Cushman Laboratory for Foraminiferal Research, 1 (3):7, pl.1, fig. 9a-c.

Type locality Palacho Hacienda, south of Panuco-Tampico R.R., State of Vera Cruz, Mexico.

Diagnosis Test low trochospiral, generally plano-convex, umbilical side strongly convex; equatorial periphery lobulate, axial periphery rounded in early portion to angular in last portion. Wall perforate, surface prominently spinose throughout.

Chambers inflated, early chambers moderately compressed, subglobular to subangular, last chambers strongly compressed, subangular to acute, arranged in about three whorls; the four chambers of the last whorl increase rapidly in size, showing on spiral side a kind of imbricate structure.

Sutures on spiral side curved, depressed; on umbilical side radial, depressed.

Umbilicus narrow, deep.

Primary aperture a medium, interiomarginal, extraumbilical-umbilical arch, bordered by a rim, with one or more single secondary sutural apertures visible on the spiral side at the inner margin of the later chambers where they lie against the previous whorl.

Strat. distr. Upper part of *Globigerapsis kugleri* zone to top of *Orbulinoides beckmanni* zone. Questionable occurrence in lower part of *Globigerapsis kugleri* zone.

Remarks Locality of figured specimen is Hg 8581, Trinidad.

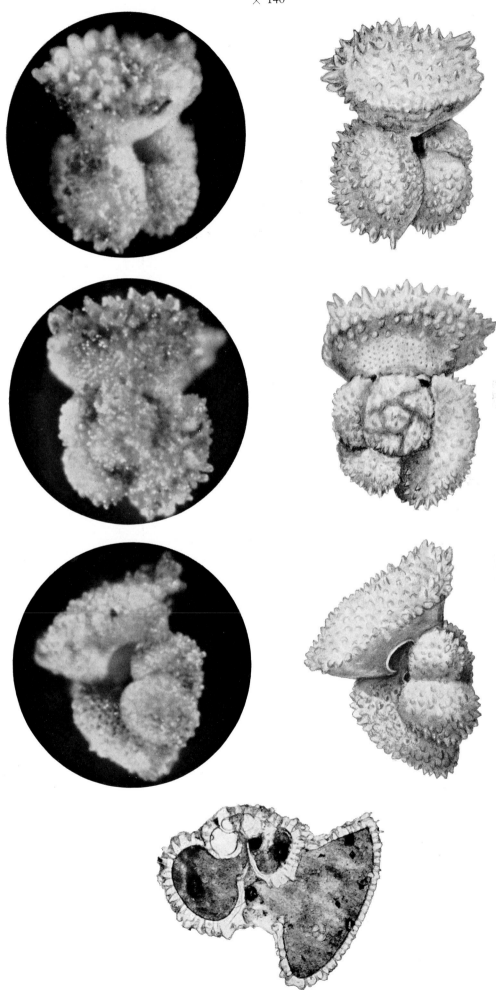

Figure 13 — Foraminiferal pack wackestone with *Globigerina triloculinoides* PLUMMER, *Globorotalia pseudobulloides* (PLUMMER) and *Globorotalia compressa* (PLUMMER).

× 30

Sample Bn 113 of a section between Probbico and San Lorenzo, Marches-Umbria area, Italy.

Probably *Globigerina daubjergensis* zone.

Figure 14 — Foraminiferal wackestone with *Globigerina triloculinoides* PLUMMER, *Globorotalia angulata* (WHITE) and *Globorotalia ehrenbergi* BOLLI. × 25

Sample Bn 29, Marches-Umbria area, Italy.

Globorotalia angulata zone.

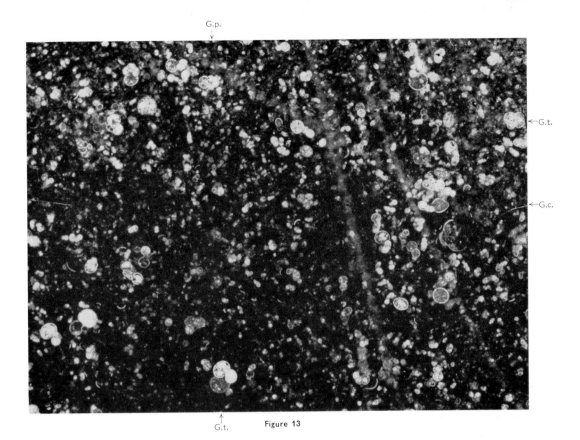

G.p.

←G.t.

←G.c.

G.t.

Figure 13

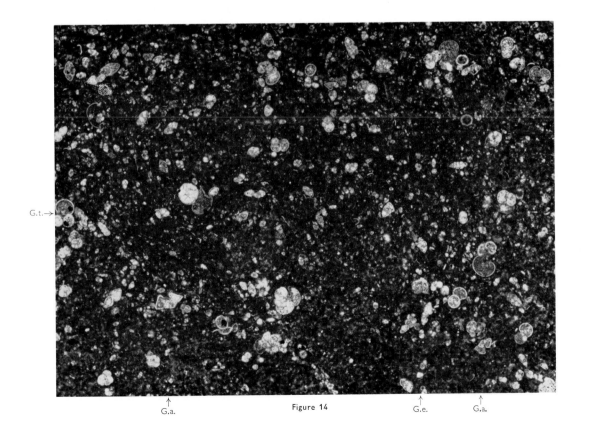

G.t.→

G.a.

Figure 14

G.e.

G.a.

238

Figure 15 — Foraminiferal wackestone with *Globigerina linaperta* FINLAY, *Globorotalia pseudomenardii* BOLLI and *Globorotalia laevigata* BOLLI. × 33
Sample Bn 386 of the Fossombrone section, Marches-Umbria area, Italy.
Globorotalia pseudomenardii zone.

Figure 16 — Foraminiferal packstone with *Globorotalia velascoensis* (CUSHMAN). × 30
Sample Ke 685 of the Khude Range section, West Pakistan.
Probably *Globorotalia velascoensis* zone.

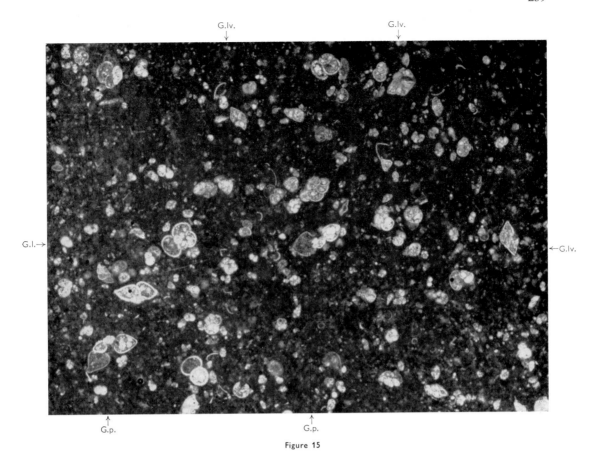

Figure 15

Figure 16

240

Figure 17 — Foraminiferal wackestone with *Globigerapsis kugleri* Bolli, Loeblich and Tappan and *Truncorotaloides topilensis* (Cushman). × 30
Sample Bn 256 of the Gubbio section, Marches-Umbria area, Italy.
Upper part *Globigerapsis kugleri* zone – *Orbulinoides beckmanni* zone.

Figure 18 — Foraminiferal wackestone with *Globorotalia cerroazulensis* (Cole), *Globorotalia centralis* Cushman and Bermudez, *Hantkenina alabamensis* Cushman and *Hantkenina* sp. × 17
Sample Ke 806 of the Zarro Range section, West Pakistan.
Globigerapsis mexicana zone – *Globorotalia cerroazulensis* zone.

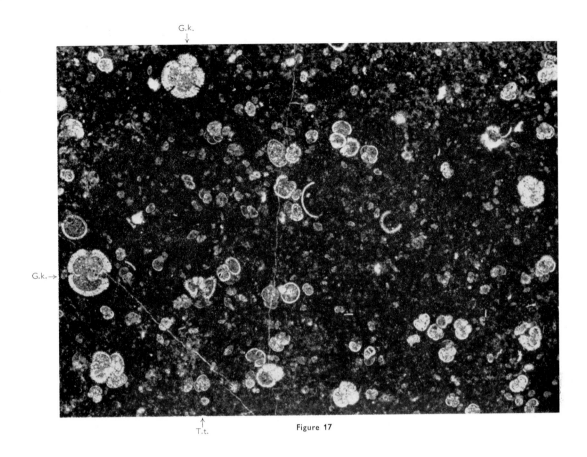

Figure 17

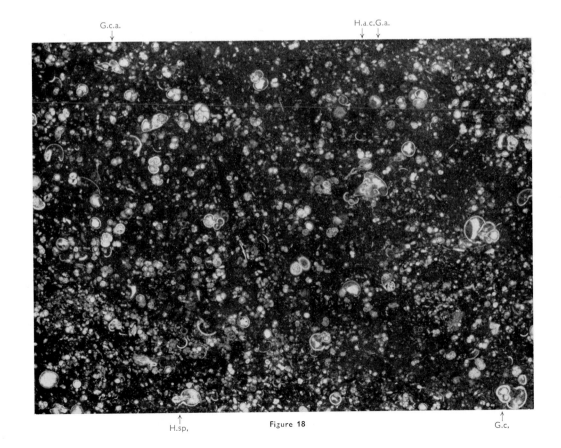

Figure 18

Oligocene-Quaternary

DESCRIPTIONS AND ILLUSTRATIONS OF
OLIGOCENE-QUATERNARY SPECIES
(arranged in alphabetical order)

Candeina nitida D'ORBIGNY

Reference — *Candeina nitida* D'ORBIGNY, 1839: Foraminifères. In: DE LA SAGRA, Histoire physique, politique et naturelle de l'Ile de Cuba. — 8:107, pl.2, figs. 27-28.

Type locality — Not designated. Localities given: Cuba and Jamaica.

Diagnosis — Test medium high trochospiral; equatorial periphery lobulate; axial periphery broadly rounded.
Wall finely perforate; surface smooth.
Chambers globular to hemispherical, arranged in about four whorls; increasing regularly in size, except for the last chamber, which may be somewhat smaller; the last whorl consists of three chambers.
Sutures slightly curved to radial, depressed.
No primary aperture in the adult test; small rounded sutural secondary apertures, bordered by a rim, almost completely surround most later chambers.

Strat. distr. — Upper part *Globorotalia acostaensis* zone to Recent. Questionable occurrence in lower part *G. acostaensis* zone.

Remarks — Locality of figured specimen is 516 "Downwind", BG 121, lat. 27° 09' S., long. 109° 50' W., Pacific Ocean.

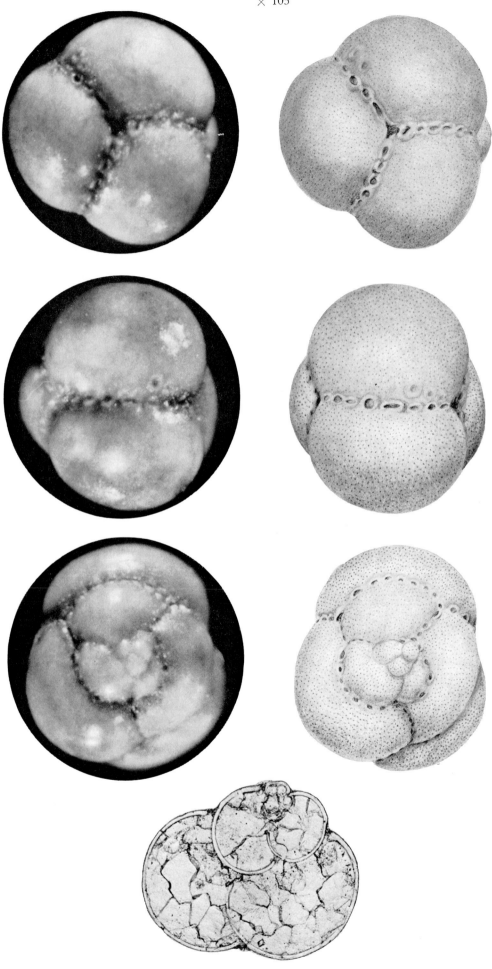

Cassigerinella chipolensis (CUSHMAN and PONTON)

Reference   *Cassidulina chipolensis* CUSHMAN and PONTON, 1932: The foraminifera of the Upper, Middle and part of the Lower Miocene of Florida. — Florida Geol. Surv. Bull., 9:98, pl. 15, figs. 2a-c.

Type locality   Locality 19, at mouth of Senterfeit Branch, Chipola River, Calhoun County, Florida, U.S.A

Diagnosis   Test small, in early stage planispiral, later with biserially arranged chambers continuing to spiral in the same plane; biumbilicate; equatorial periphery lobulate, axial periphery rounded.
Wall perforate, surface smooth.
Chambers in the early stage globular to subglobular, those of later stage subglobular to ovate.
Sutures slightly curved, depressed.
Aperture interiomarginal, extraumbilical, an elongate, loop-shaped slit, alternating in position from one side to the next in successive chambers.

Strat. distr.   Base of *Globigerina ampliapertura* zone into upper part of *Globorotalia siakensis* zone. Questionable occurrence in uppermost part of *G. siakensis* zone and lowermost part of *Globorotalia menardii* zone.

Remarks   It is doubtful whether this species belongs to the group of planktonic Foraminifera in view of the imperforate wall structure.
Locality of figured specimen is the type locality of the *Catapsydrax dissimilis* zone, Trinidad, W.I.

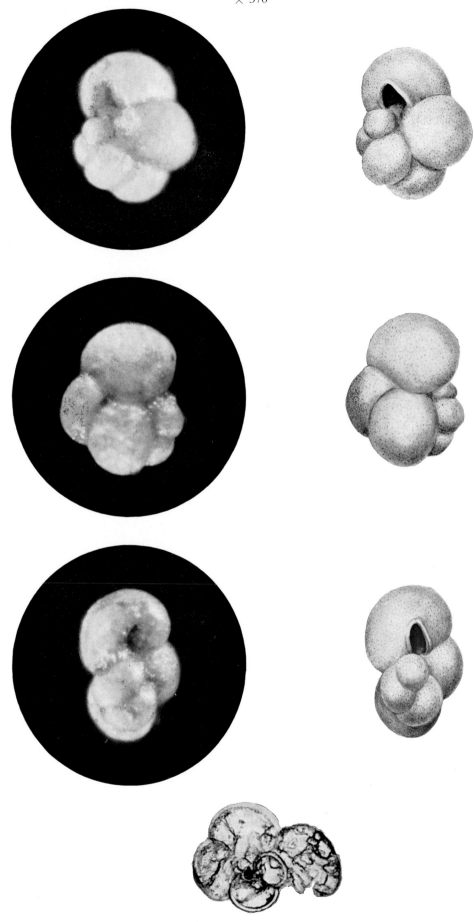

Catapsydrax dissimilis (Cushman and Bermudez)

Reference   *Globigerina dissimilis* Cushman and Bermudez, 1937: Further new species of Foraminifera from the Eocene of Cuba. — Contributions from the Cushman Laboratory for Foraminiferal Research, 13 (1):25, pl. 3, figs. 4-6.

Type locality   Bermudez Station 31,1 km. N. of Arroyo Arenas, on road to Jaimanitas (water well), Habana Province, Cuba.

Diagnosis   Test low to medium trochospiral, spiral side slightly convex, umbilical side strongly convex; equatorial periphery lobulate; axial periphery broadly rounded.
Wall distinctly perforate, surface pitted.
Chambers spherical, last chambers somewhat laterally compressed, arranged in about three whorls; the four chambers of the last whorl increase rapidly in size; the final chamber may be slightly reduced in size.
Sutures on spiral side curved to slightly curved, depressed; on umbilical side radial, depressed.
Primary aperture interiomarginal, umbilical, in the final stage covered by a single umbilical bulla, with 2-4 accessory infralaminal apertures.

Strat. distr.   Upper part *Globigerina angulisuturalis* zone to top of *Globigerinoides trilobus* zone. *Catapsydrax* cf. *dissimilis* ranges throughout the *Globigerina ampliapertura* zone and the lower part of the *G. angulisuturalis* zone.

Remarks   *Catapsydrax* cf. *dissimilis* has 2 instead of 3–4 accessory infralaminal apertures. See remarks *Catapsydrax stainforthi* Bolli, Loeblich and Tappan.
Locality of figured specimen is DB 83, Trinidad, W.I.

Erratum

Diagnosis, last line. Read "3-4 accessory infralaminal apertures" in stead of "2-4 accessory infralaminal apertures".

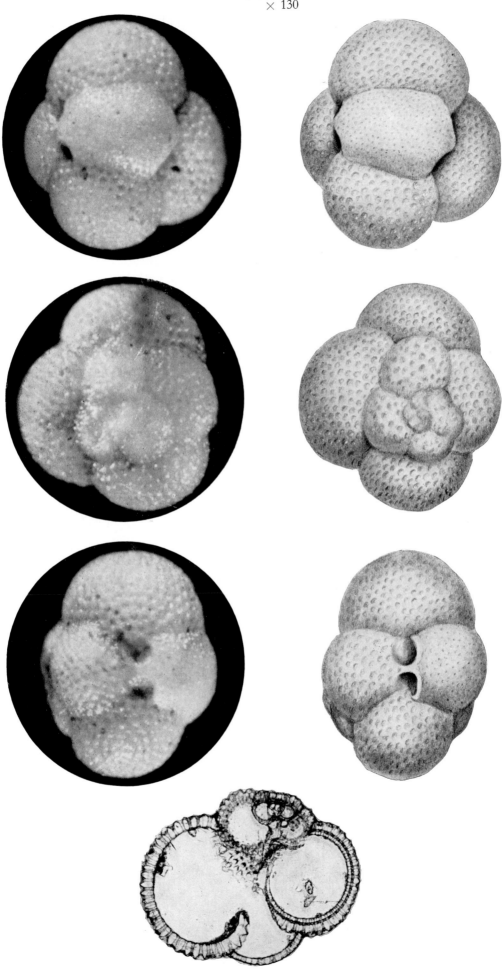

Catapsydrax stainforthi BOLLI, LOEBLICH and TAPPAN

Reference   *Catapsydrax stainforthi* BOLLI, LOEBLICH and TAPPAN, 1957: Planktonic foraminiferal families Hantkeninidae, Orbulinidae, Globorotaliidae and Globotruncanidae. — United States National Museum Bulletin, 215:37, pl. 7, figs. 11a-c.

Type locality   A surface from a section along the Cipero coast, Cipero Formation, western Trinidad, W.I.

Diagnosis   Test low to medium trochospiral; equatorial periphery lobulate; axial periphery rounded. Wall distinctly perforate, surface pitted.
Chambers spherical to subglobular, arranged in about three and a half whorls; the four or occasionally five chambers of the last whorl increase regularly in size, the final chamber may be slightly reduced in size.
Sutures radial to slightly curved, depressed.
Primary aperture interiomarginal, umbilical, in the final stage covered by a single umbilical bulla with a small infralaminal accessory opening over each suture of the final whorl.

Strat. distr.   Lower part of *Globigerinoides trilobus* zone to upper part of *Globigerinatella insueta* zone.

Remarks   This species differs from *Catapsydrax dissimilis* (CUSHMAN and BERMUDEZ) in being smaller and in having a more closely appressed bulla which usually extends a short way along the sutures and has smaller arched accessory openings that are restricted to the area over the sutures.
Locality of figured specimen is the type locality of the *Catapsydrax stainforthi* zone, Trinidad, W.I.

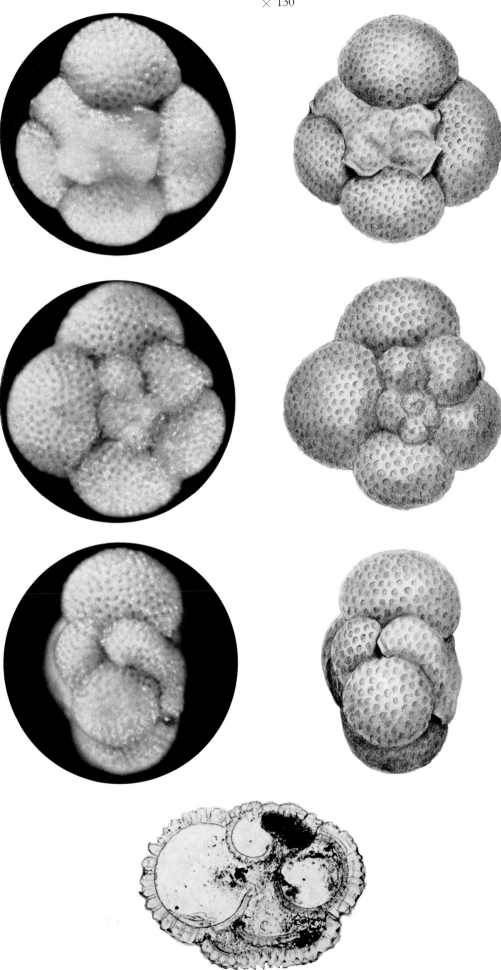

Globigerina angulisuturalis BOLLI

Reference   *Globigerina ciperoensis angulisuturalis* BOLLI, 1957: Planktonic Foraminifera from the Oligocene-Miocene Cipero and Lengua Formations of Trinidad, B.W.I. — United States National Museum Bulletin, 215:109, pl. 22, figs. 11a-c.

Type locality   Type locality of the *Globorotalia opima* zone, Trinidad, Sample Bo 306A (TTOC 215657).

Diagnosis   Test low to medium trochospiral; equatorial periphery almost circular, lobulate, with deep, angular U-shaped sutures between the chambers; axial periphery rounded.
Wall perforate, surface finely pitted, may be slightly rugose near shoulders of early chambers of last whorl.
Chambers spherical, arranged in about three and a half whorls; the five chambers of the last whorl increase moderately in size.
Sutures on spiral side broadly depressed, radial; on umbilical side broadly depressed, radial.
Umbilicus wide, fairly deep.
Aperture interiomarginal, umbilical, with a fairly high arch, bordered by a faint rim.

Strat. distr.   Ranging throughout the *Globigerina angulisuturalis* zone and the lower part of the *Globorotalia kugleri* zone.

Remarks   *Globigerina angulisuturalis* BOLLI is distinguished from *G. ciperoensis* BOLLI by having deep cut, angular U-shaped sutures.
Locality of figured specimen is the type locality of the *Globigerina ciperoensis* zone, Trinidad, W.I.

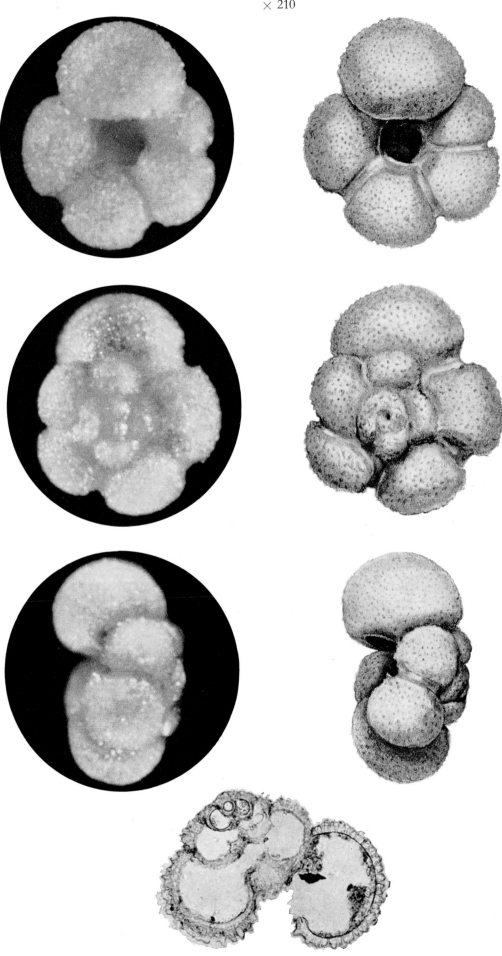

Globigerina binaiensis Koch

Reference
Globigerina? aspera Koch, 1926: Mitteltertiäre Foraminiferen aus Bulongan, Ost-Borneo. — Eclogae Geologicae Helvetiae, 19 (3):746, figs. 22a-c.
Globigerina binaiensis Koch, 1935: Namensänderung einiger Tertiär-Foraminiferen aus Niederländisch Ost-Indien. — Eclogae Geologicae Helvetiae, 28 (2):558.

Type locality
Not designated. Localities given: Sadjau-Njak, Binai-Atingdunok 491 and Binai-Atingdunok 504, S.E. Bulongan, East Kalimantan, Indonesia.

Diagnosis
Test low trochospiral; equatorial periphery subcircular to subquadrate, slightly lobulate; axial periphery broadly rounded, with a dorso-peripheral shoulder.
Wall distinctly perforate, except on characteristic apertural face, surface pitted and may be hispid.
Chambers spherical; the final chamber laterally compressed, arranged in about three and a half whorls; the three chambers of the last whorl increase rapidly in size.
Sutures on spiral side slightly curved to radial, depressed; on umbilical side radial, depressed.
Umbilicus fairly small.
Aperture is a fairly low arch, interiomarginal, umbilical, extends the width of the umbilicus at the base of the apertural face, which is bounded by lateral lobe-like expansions of the last chamber. A lip, wich may broaden medially to form a weak "umbilical tooth", is present.

Strat. distr.
Ranging throughout upper part *Globigerina angulisuturalis* zone, *Globorotalia kugleri* zone and lower part *Globigerinoides trilobus* zone. Questionable occurrence in middle part *G. angulisuturalis* zone.

Remarks
This species is distinguished from *Globigerina sellii* (Borsetti) by its typically high apertural face, the surface of which is almost smooth and non-perforate, and the acute shape of the last chamber.
See remarks *Globigerina sellii* (Borsetti).
Locality of figured specimen is H 280a (4m), E.Java, Indonesia.

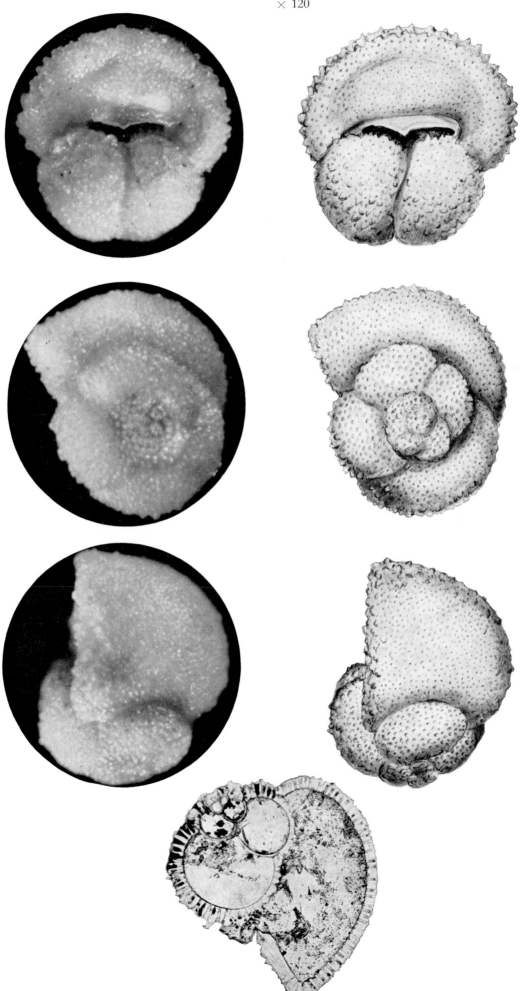

Globigerina ciperoensis BOLLI

Reference *Globigerina ciperoensis* BOLLI, 1954: Note on *Globigerina concinna* REUSS 1850. — Contributions from the Cushman Foundation for Foraminiferal Research, 5 (1):1.

Type locality Northern end of the Cipero Coast, south of San Fernando, Trinidad.

Diagnosis Test low to medium trochospiral; equatorial periphery almost circular, lobulate; axial periphery rounded.
Wall perforate, surface fairly pitted, may be slightly rugose near shoulders of early chambers of last whorl.
Chambers spherical, arranged in about three and a half whorls; the five chambers of the last whorl increase moderately to rapidly in size.
Sutures on spiral side depressed, radial; on umbilical side depressed, radial.
Umbilicus wide, radial.
Aperture interiomarginal, umbilical, with a fairly high arch, bordered by a faint rim.

Strat. distr. Lower part of *Globigerina ampliapertura* zone to top of *Globigerina angulisuturalis* zone. Questionable occurrence in lowermost part of *G. ampliapertura* zone and lowermost part of *Globorotalia kugleri* zone.

Remarks In the literature it appears that this species is described many times under the name *Globigerina* cf. *concinna* REUSS. REUSS' types are Tortonian (Upper Miocene) in age, and unfortunately have been lost. Direct comparison is thus not possible. See remarks *Globigerina angulisuturalis* BOLLI.
Locality of figured specimen is the above mentioned type locality.

× 190

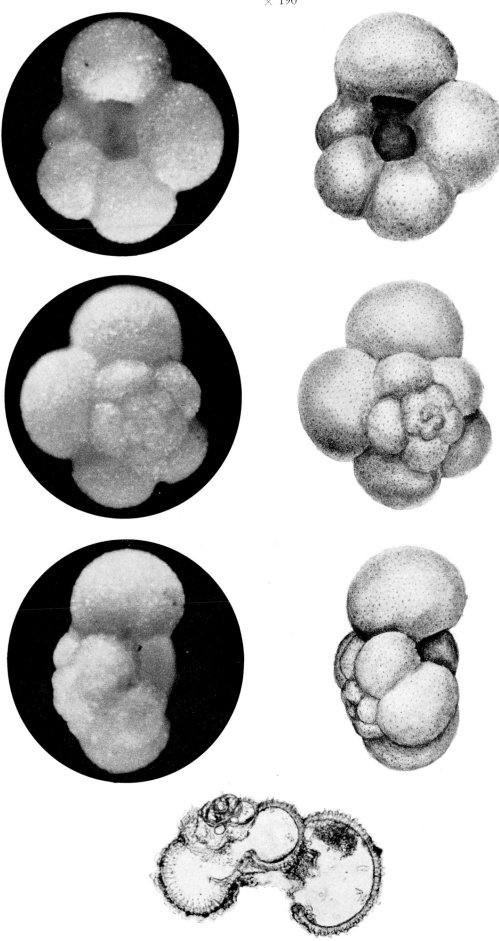

Globigerina nepenthes TODD

Reference   *Globigerina nepenthes* TODD, 1957: Geology of Saipan, Mariana Islands; Part 3 — Paleontology. — U.S. Geol. Survey Prof. Paper, 280-H:301, pl.78, figs. 7a-b.

Type locality   Locality S 621, east-central Saipan, Mariana Islands.

Diagnosis   Test low trochospiral, compact; equatorial periphery slightly lobulate except for the last part; axial periphery rounded.
Wall distinctly perforate, surface pitted with slight rugosities near umbilical area.
Chambers inflated, the last one strongly protruding, arranged in about three and a half whorls, with four to five chambers in the last whorl.
Sutures on spiral side slightly curved, slightly depressed; on umbilical side almost radial, depressed.
Umbilicus narrow, shallow.
Aperture interiomarginal, umbilical, a semicircular arch, bordered by a slightly upturned, distinct rim.

Strat. distr.   Upper part *Globorotalia siakensis* zone into lower part *Globoquadrina altispira* zone. Questionable occurrence in lower part *G. siakensis* zone.

Remarks   Locality of figured specimen is A.H. 3692, Trinidad, W.I.

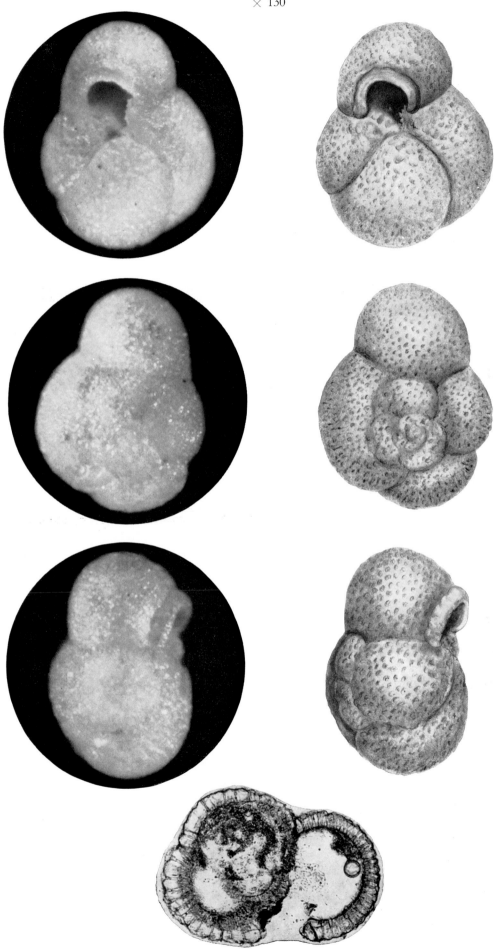

Globigerina praebulloides Blow

Reference *Globigerina praebulloides* Blow, 1959: Age, correlation and biostratigraphy of the Upper Tocuyo (San Lorenzo) and Pozón Formations, eastern Falcón, Venezuela. – Bulletins of American Paleontology, 39 (178): 180, pl. 8, figs. 47 a-c.

Type locality Holotype from an auger line near the north to south section of the Pozón – El Mene Road between Caiman and Buena Vista, near Pozón, eastern Falcón, Venezuela.

Diagnosis Test low trochospiral; equatorial periphery distinctly lobulate; axial periphery rounded. Wall perforate, surface pitted.
Chambers inflated, slightly embracing, arranged in about two and a half whorls with four chambers in the last whorl, increasing fairly rapidly in size.
Sutures on spiral side radial to slightly curved, depressed; on umbilical side radial, depressed.
Umbilicus narrow.
Aperture interiomarginal, umbilical, a low to moderate arch, bordered by a weak rim.

Strat. distr. Base *Globigerina ampliapertura* zone to upper part *Globorotalia dutertrei* zone. Questionable occurrence in uppermost Eocene.

Remarks *Globigerina praebulloides* Blow is regarded as the ancestor of *Globigerinoides primordius* Blow.
Locality of figured specimen is DB 288, Trinidad, W.I.

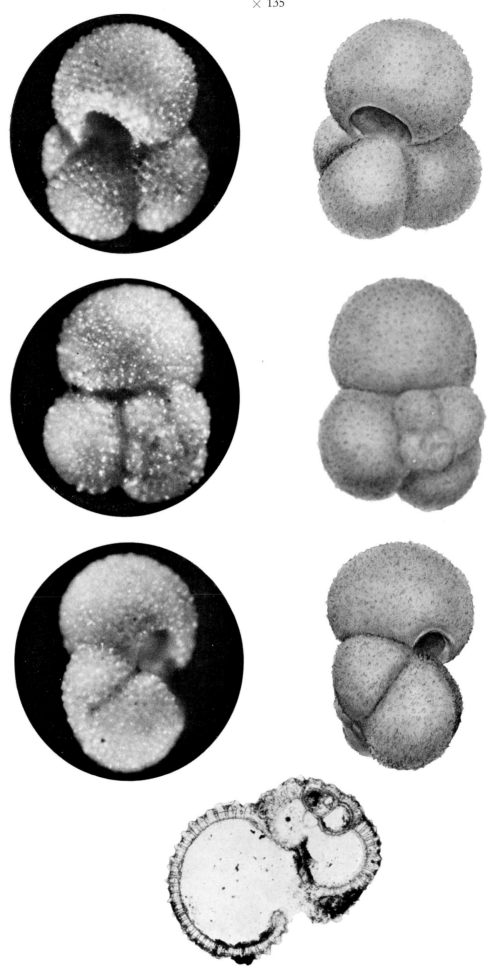

Globigerina riveroae BOLLI and BERMUDEZ

Reference   *Globigerina riveroae* BOLLI and BERMUDEZ, 1965: Zonation based on Planktonic Foraminifera of Middle Miocene to Pliocene warm-water sediments. — Bol. Inf. Asoc. Ven. Geol., Min. Petr., 8(5):137, pl.1, figs. 1-6.

Type locality   Outcrop directly below the terrace on which stands the village of Araya, Peninsula de Araya, Venezuela. Sample P. J. Bermudez 7/64.

Diagnosis   Test low to medium high trochospiral, wider than high; equatorial periphery strongly lobulate; axial periphery rounded.
Wall perforate; surface slightly pitted.
Chambers almost spherical, arranged in about three and a half whorls; the four chambers of the last whorl increase rapidly in size.
Sutures on spiral and umbilical side almost radial, depressed.
Umbilicus very wide, deep.
Aperture interiomarginal, umbilical, a very large, more or less semicircular, arch bordered by a distinct though thin rim. Aperture of penultimate chamber also clearly visible.

Strat. distr.   Upper part *Globorotalia margaritae* zone to top *Globoquadrina altispira* zone.

Remarks   This species differs from *Globigerina bulloides* D'ORBIGNY in having a larger aperture. Locality of figured specimen is well Cubagua No. 1, core sample 734-739 feet, Venezuela.

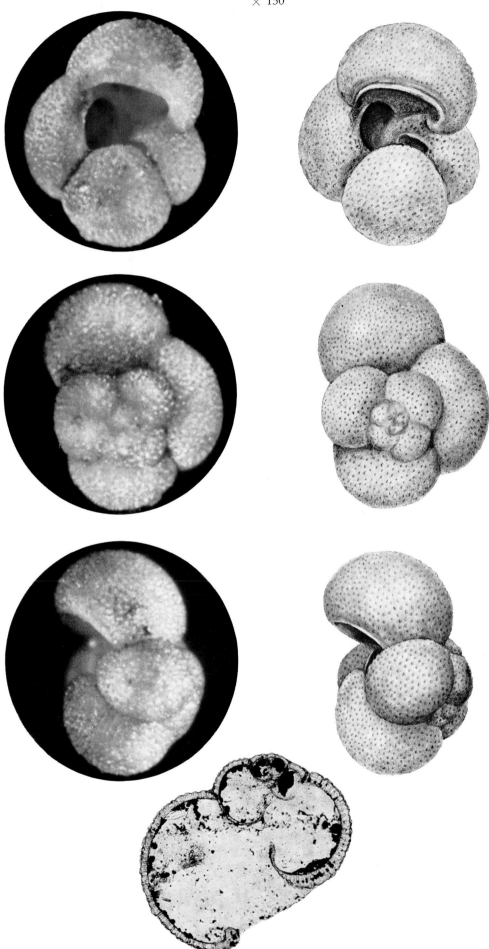

Globigerina sellii (Borsetti)

Reference *Globoquadrina sellii* Borsetti, 1959: Tre nuovi Foraminiferi Planctonici dell' Oligocene Piacentino. —· Giornale di Geologia, Annali del Museo Geologico di Bologna, XXVII: 209, pl. XIII, figs. 3a-d.

Type locality 1.5 km west of Vigoleno, along the path which connects the Varani area with the Vigoleno-Vernasca road, northern Italy.

Diagnosis Test low trochospiral; equatorial periphery subcircular to subquadrate, slightly lobulate; axial periphery broadly rounded, subconical, with a dorso-peripheral shoulder.
Wall distinctly perforate, surface pitted and may be hispid.
Chambers spherical, those of last whorl laterally compressed, reniform, arranged in about three and a half whorls; the three chambers of the last whorl increase rapidly in size.
Sutures on spiral side slightly curved to radial, depressed; on umbilical side radial, depressed.
Umbilicus fairly small, triangular, deep.
Aperture is a fairly low arch, interiomarginal, umbilical, extends the width of the umbilicus at the base of the apertural face, which is bounded by lateral lobe-like expansions of the last chamber. A lip, which may broaden medially to form a weak "umbilical tooth", is present.

Strat. distr. Upper part *Globigerina ampliapertura* zone and *Globigerina angulisuturalis* zone. Questionable occurrence in lower part *Globigerina ampliapertura* zone and lower part *Globorotalia kugleri* zone.

Remarks In our opinion *Globigerina sellii* (Borsetti) and *G. binaiensis* Koch are closely related to *G. rohri* Bolli and the *G. tripartita* Koch group. When the last chamber is removed from specimens of this latter group a striking resemblance with *G. sellii* or *G. binaiensis* is observed. In Trinidad, specimens of the *G. tripartita* group (U.S.N.M. Bulletin 215, pl. 23, figs. 4a-b) occur in great quantity, while specimens of *G. sellii*, although not so numerous, are also known from the same Trinidad samples. In East Java, Indonesia, *G. tripartita, G. sellii* and *G. binaiensis* are quite common. Specimens of the latter two species, showing remnants of an additional chamber, have been observed, giving the impression that *G. sellii* and *G. binaiensis* are pre-adult stages of the *G. tripartita/rohri* group.
The generic position of these species is in some doubt. It is likely that they are links in a possible *Globigerina yeguaensis-Globoquadrina dehiscens* lineage.
Globigerina oligocaenica Blow and Banner, 1957 is a synonym of *Globigerina sellii*.
Crescenti (1966) suggests that these two species are synonymous with *Globoquadrina obesa* Akers, 1955.
Locality of figured specimen is FCRM 1627, Lindi, E.Africa.

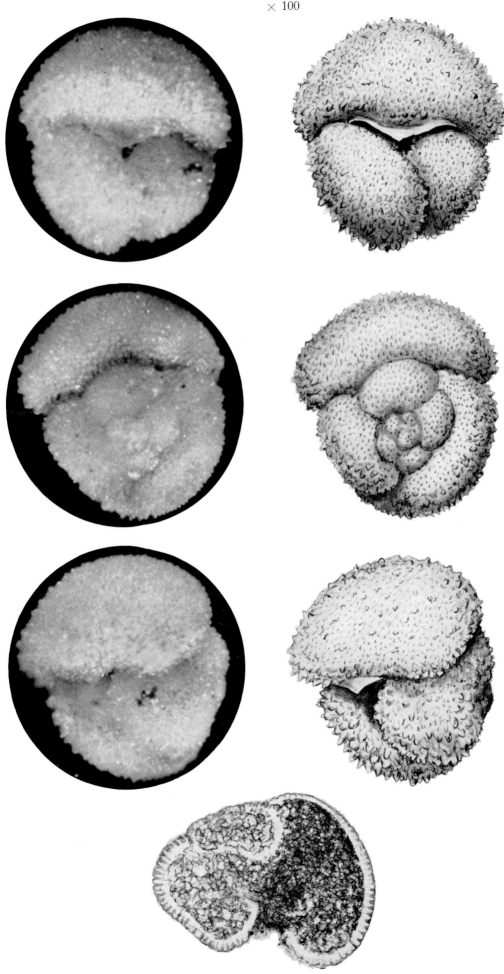

Globigerina seminulina SCHWAGER

Reference *Globigerina seminulina* SCHWAGER, 1866: Fossile Foraminiferen von Kar Nikobar. Novara Exped. 1857-1859. — Geol. Theil, 2 (2):256.

Type locality Kar Nikobar, India.

Diagnosis Test low trochospiral; equatorial periphery lobulate; axial periphery rounded to subangular in the last chamber.
Wall thick, distinctly perforate, surface pitted with rugosities near umbilical edge or almost smooth.
Chambers spherical, the last one most often elongate, arranged in about three and a half whorls, with three to five chambers in the last whorl.
Sutures on spiral side slightly curved to radial, depressed; on umbilical side radial, depressed.
Umbilicus small to fairly wide, fairly deep.
Aperture is an elongate slit or low arch, interiomarginal, umbilical, bordered by a rim.

Strat. distr. Lower part of *Globorotalia peripheroronda* zone into lower part of *Globoquadrina altispira* zone. Questionable occurrences in lowermost part of *G. peripheroronda* zone, uppermost part of *Globigerinatella insueta* zone and middle part of *G. altispira* zone.

Remarks *Globigerina seminulina* SCHWAGER is a variable species. As a result of this a number of synonyms have been published as new species.
They are as follows:
Globigerina kochi CAUDRI, 1934
Sphaeroidinella disjuncta FINLAY, 1940
Sphaeroidinella rutschi CUSHMAN and RENZ, 1941
Globigerina grimsdalei KEYZER, 1945
The most variable features are the wall structure, the number of chambers in the last whorl and the shape of the last chamber. The wall of the chambers may be greatly thickened and sometimes a kind of smooth, translucent coating (cortex) may be partly present. The final chamber may be of smaller than normal size, or elongate sack-like in several gradations. According to these features it is easy to split *Globigerina seminulina* into several "subspecies". It should be determined locally whether such a splitting has stratigraphic usefulness.
For some views on the validity of the genus *Sphaeroidinella* reference should be made to A.W.H.Bé, Micropaleontology, 11 (1):81.
Locality of figured specimen of the type locality of the *Globorotalia robusta* zone, Trinidad, W.I.

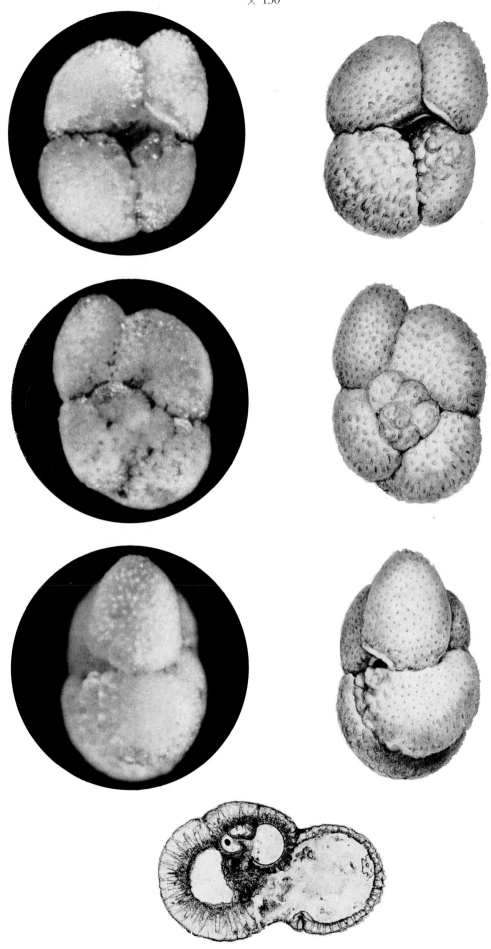

Globigerina tripartita Koch

Reference *Globigerina bulloides* D'Orbigny, var. *tripartita* Koch, 1926: Mitteltertiäre Foraminiferen aus Bulongan, Ost-Borneo. – Eclogae Geologicae Helvetiae XIX (3): 746, figs. 21a-b.

Type locality Not designated. Localities given: Sadjau-Nak and Binai-Atingdunok 491, S. E. Bulongan, E. Borneo, Indonesia.

Diagnosis Test low to medium trochospiral; equatorial periphery slightly lobulate.
Wall thick, distinctly perforate, surface smooth, except for the umbilical edge, which has rugosities or short, thick spines.
Chambers spherical, those of last whorl strongly laterally compressed, arranged in about three and a half whorls; the three chambers of the last whorl increase very rapidly in size; in large specimens the final chamber is usually reduced in size.
Sutures on spiral side curved, depressed; on umbilical side radial, depressed.
Umbilicus fairly narrow to fairly wide, triangular.
Aperture is an interiomarginal, umbilical arch, deep in umbilicus.

Strat. distr. Base *Globigerina ampliapertura* zone to upper part *Globigerinoides trilobus* zone. Questionable occurrence in the Upper Eocene.

Remarks This species is distinguished from *Globigerina venezuelana* Hedberg by having three instead of four chambers in the last whorl and by having the chambers of the last whorl laterally more compressed. See remarks *Globigerina sellii* (Borsetti).
Locality of figured specimen is DB 93, Trinidad, W.I.

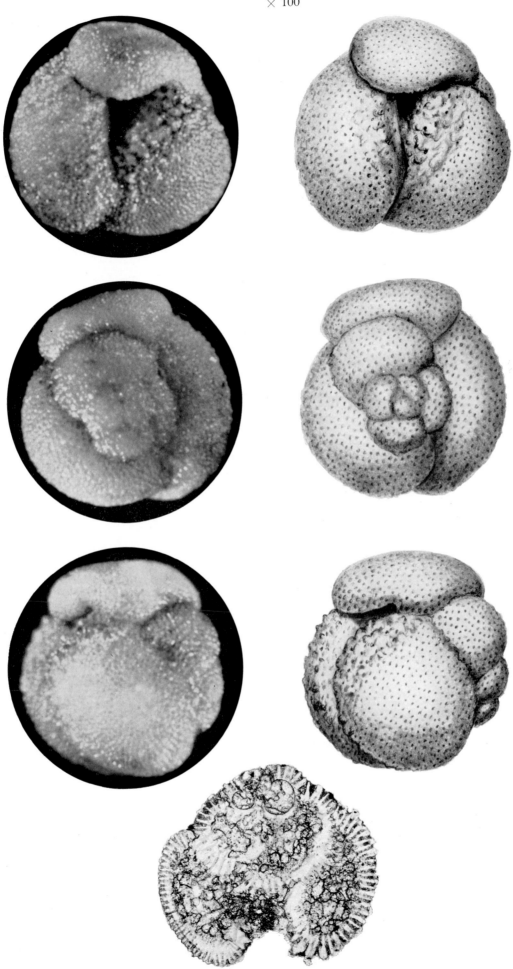

Globigerina venezuelana HEDBERG

Reference *Globigerina venezuelana* HEDBERG, 1937: Foraminifera of the Middle Tertiary Carapita Formation of northeastern Venezuela. — Journal of Paleontology, 11 (8):681, pl. 92, figs. 7a-b.

Type locality North of the town of Santa Ines on Quebrada Carapita, a small tributary of the Rio Querecal, District of Libertad, State of Anzoategui, northeastern Venezuela.

Diagnosis Test low to medium trochospiral; equatorial periphery slightly lobulate.
Wall thick, distinctly perforate, surface smooth, except for the umbilical edge, which has rugosities or short, thick spines.
Chambers spherical, those of last whorl laterally compressed, arranged in about three and a half whorls; the four chambers of the last whorl increase rapidly in size; the final chamber is commonly reduced in size.
Sutures on spiral side curved to slightly curved, depressed; on umbilical side radial, depressed.
Umbilicus fairly narrow to fairly wide, triangular in normal specimens.
Aperture is an interiomarginal, umbilical arch, deep in umbilicus.

Strat. distr. Lower part *Globigerina angulisuturalis* zone into lower part *Globoquadrina altispira* zone. Questionable occurrence in lowermost part *G. angulisuturalis* zone.

Remarks *Globigerina venezuelana* HEDBERG might be a synonym of *Globigerina quadripartita* KOCH. Locality of figured specimen is PM 1924, Trinidad, W.I.

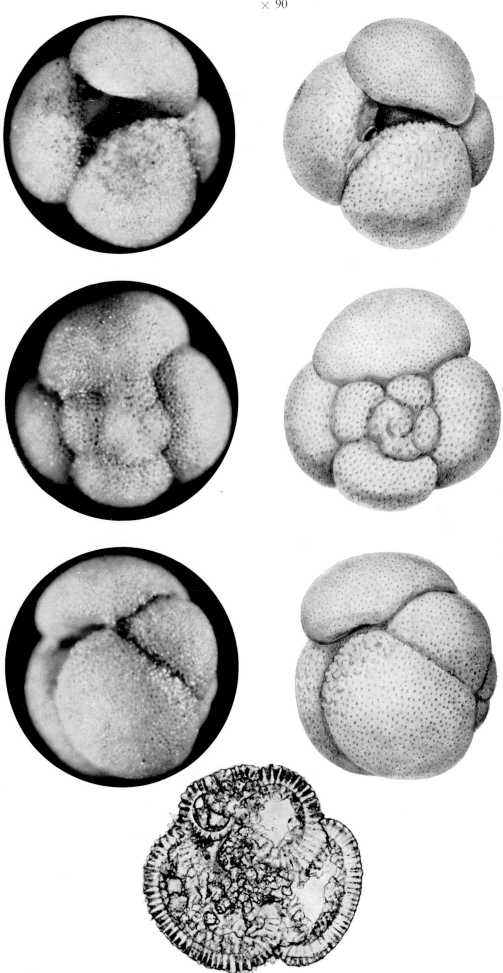

Globigerinatella insueta CUSHMAN and STAINFORTH

Reference *Globigerinatella insueta* CUSHMAN and STAINFORTH, 1945: The foraminifera of the Cipero marl formation of Trinidad, British West Indies. — Cushman Laboratory for Foraminiferal Research, Special Publ., 14:69, pl. 13, figs. 7-9.

Type locality Sample Rz. 108, Trinidad Leaseholds Catalogue No. 21743, along the coast between the mouth of the Cipero River and the point at which the Trinidad Government railway turns inland, south of San Fernando, Trinidad, W.I.

Diagnosis Test subglobular, early portion trochospiral with the final chamber embracing.
Wall distinctly perforate, surface pitted.
Chambers of the early portion (*Globigerina* stage) spherical, rapidly increasing in size; the large enveloping final chamber constitutes a considerable part of the test.
Sutures radial, depressed.
Aperture in the early stage interiomarginal, umbilical; in the adult stage with secondary sutural and areal apertures, surrounded by distinct lips, with small knobby pustule-like bullae covering the areal secondary apertures. More irregular, spreading, sutural bullae covering the secondary sutural apertures may be present; all bullae may have infralaminal accessory apertures.

Strat. distr. Upper part of *Globigerinoides trilobus* zone into lowermost part of *Globorotalia peripheroronda* zone.

Remarks Locality of figured specimen is AGH 2695, Trinidad, W.I.

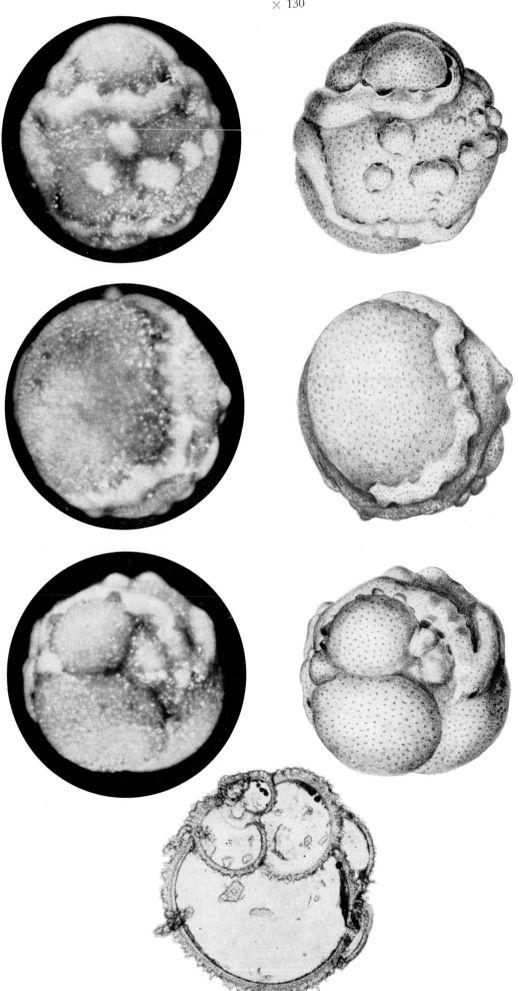

Globigerinita naparimaensis BRONNIMANN

Reference *Globigerinita naparimaensis* BRONNIMANN, 1951: *Globigerinita naparimaensis* n.gen., n.sp., from the Miocene of Trinidad, B.W.I. — Contributions from the Cushman Foundation for Foraminiferal Research, 2 (1):18.

Type locality Trinidad Leaseholds Ltd. sample no. 161214, Naparima area, Trinidad, W.I.

Diagnosis Test low to medium trochospiral; equatorial periphery lobulate, axial periphery rounded. Wall perforate, surface smooth to finely pitted, may be hispid.
Chambers spherical to subglobular, arranged in about three and a half whorls; the three to four chambers of the last whorl increase rapidly in size.
Sutures radial to slightly curved, depressed.
Primary aperture interiomarginal, umbilical, in the final stage completely covered by an irregular bulla expanding along the earlier sutures, with numerous infralaminal accessory apertures along the margins, both at the junction with the sutures of earlier chambers and along the contact with the primary chambers.

Strat. distr. Lower part of *Globigerinoides trilobus* zone to Recent.

Remarks Locality of figured specimen is East Gulf 136.

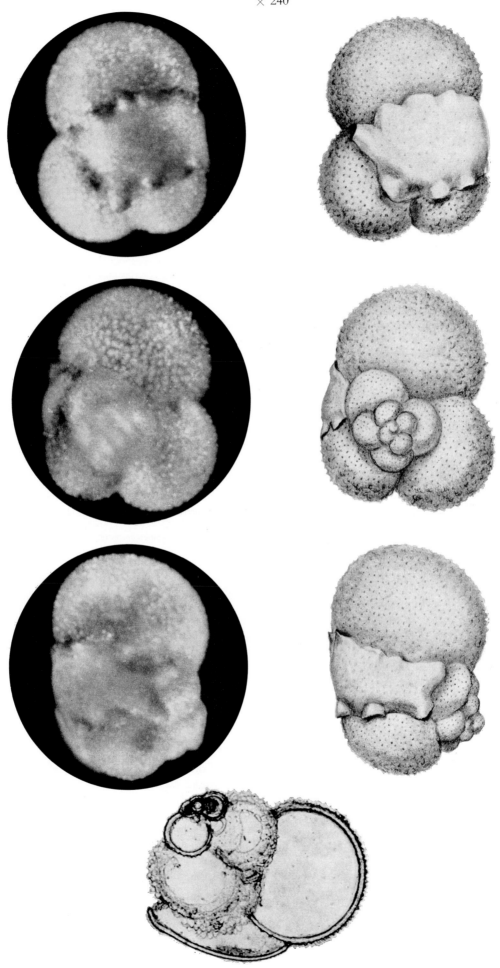

Globigerinoides altiaperturus BOLLI

Reference   *Globigerinoides triloba altiapertura* BOLLI, 1957: Planktonic Foraminifera from the Oligocene-Miocene Cipero and Lengua Formations of Trinidad, B.W.I. — United States National Museum Bulletin, 215:113, pl. 25, figs. 7a-c.

Type locality   The south bank of the San Fernando bypass road, approximately 1050 feet northeast of the north end of the road bridge across the Siparia railway line, western Trinidad, W.I.

Diagnosis   Test trochospiral, unequally biconvex; equatorial periphery distinctly lobulate; axial periphery broadly rounded.
Wall distinctly perforate, surface pitted.
Chambers spherical, arranged in about three and a half whorls; the three chambers of the last whorl increase rapidly in size.
Sutures on spiral side slightly curved, depressed; on umbilical side radial, depressed.
Umbilicus fairly narrow.
Primary aperture interiomarginal, umbilical, with a high, distinct arch, bordered by a rim; the last few chambers show a secondary sutural aperture opposite the primary one.

Strat. distr.   Lower part *Globigerinoides trilobus* zone to upper part *Globigerinatella insueta* zone. Questionable occurrence upper part *G. insueta* zone into upper part *Globorotalia siakensis* zone.

Remarks   See remarks *Globigerinoides trilobus* (REUSS).
Locality of figured specimen is the type locality of the *Catapsydrax dissimilis* zone, Trinidad, W.I.

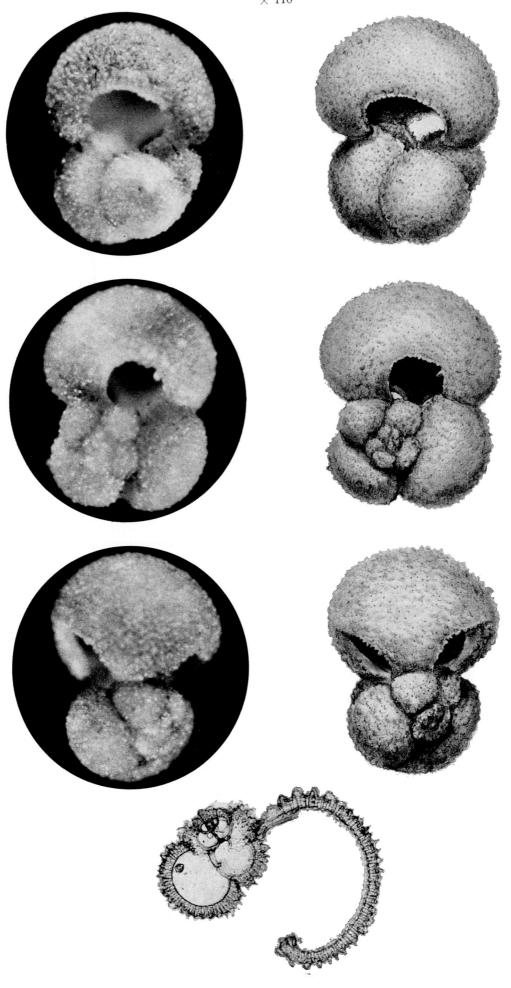

286

Globigerinoides conglobatus (BRADY)

Reference   *Globigerina conglobata* BRADY, 1879: Notes on some of the reticularian Rhizopoda of the "Challenger" Expedition; II — Additions to the knowledge of porcellanous and hyaline types. — Quart. Journ. Micr. Sci., London, n.s., 19:286. Type figure not given, see BRADY, 1884, Rept. Voy. Challenger, Zool., vol. 9, pl.80, figs. 1-5.

Type locality   Not given. Locality of figured specimen (1884): Challenger Sta. 64, lat. 35° 45' N., long. 50° 27' W., North Atlantic Ocean.

Diagnosis   Test trochospiral, subglobular to subquadrate; equatorial periphery slightly lobulate; axial periphery broadly rounded.
Wall coarsely perforate, thick; surface distinctly pitted.
Chambers in initial stage subspherical, later strongly compressed, though still inflated, arranged in about four whorls; the three to three and a half chambers of the last whorl increase slowly or hardly in size.
Sutures on spiral side obscure, on umbilical side almost radial, depressed.
Umbilicus narrow to almost closed, deep.
Primary aperture interiomarginal, umbilical, a fairly long, comparatively low, slightly asymmetric arch, bordered by a rim, the secondary sutural apertures are situated over sutures of earlier chambers, except for the last chamber, which has one extra opening.

Strat. distr.   Lower part *Globorotalia margaritae* zone to Recent. Questionable occurrence in lowermost part *G. margaritae* zone.

Remarks   Locality of figured specimen is 523 "Downwind", HG 74, lat. 28° 29' S., long. 106° 30' W., Pacific Ocean.

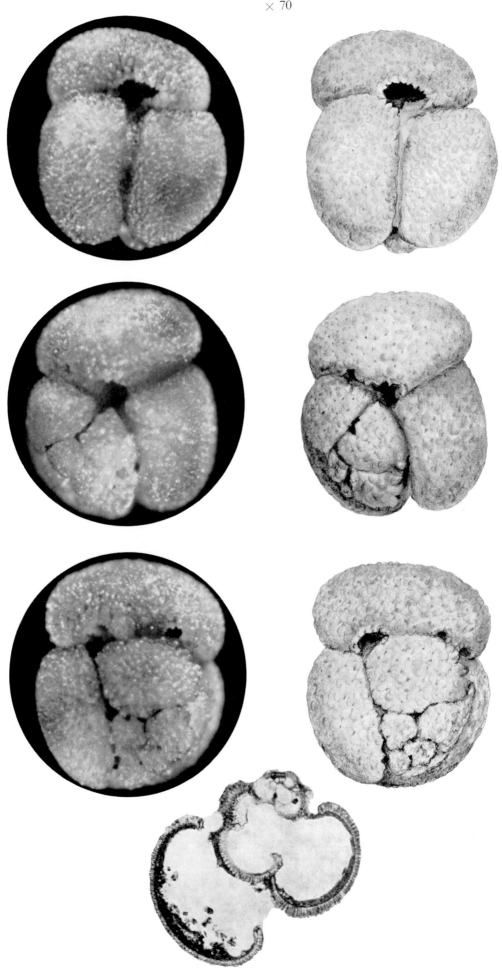

Globigerinoides diminutus BOLLI

Reference *Globigerinoides diminuta* BOLLI, 1957: Planktonic Foraminifera from the Oligocene-Miocene Cipero and Lengua Formations of Trinidad, B.W.I. — United States National Museum Bulletin, 215:114, pl. 25, figs. 11a-c.

Type locality A core at 7419-7439 feet in United British Oilfields of Trinidad, Ltd.'s well Penal No. 92, Trinidad, W.I.

Diagnosis Test trochospiral; equatorial periphery almost subquadrate; axial periphery broadly rounded.
Wall distinctly perforate, surface pitted.
Chambers spherical in early stages, later chambers somewhat laterally compressed, arranged in three and a half to four whorls; the three chambers of the last whorl increase moderately in size.
Sutures on spiral side radial to slightly curved, slightly depressed, on umbilical side radial, slightly depressed.
Umbilicus small.
Primary aperture interiomarginal, umbilical, with a small but distinct, almost circular, arch, bordered by a rim; the last few chambers show two small, but distinct, secondary sutural apertures situated over sutures of earlier chambers.

Strat. distr. Base *Globigerinatella insueta* zone into lowermost part of *Globorotalia peripheroronda* zone.

Remarks The position of the apertures of this species is the same as in *Globigerinoides ruber* (D'ORBIGNY), the test, however, is smaller and more compact.
Locality of figured specimen is RD 22, Mirimire-Tucacas Road, Falcon, Venezuela.

Globigerinoides diminutus
× 220

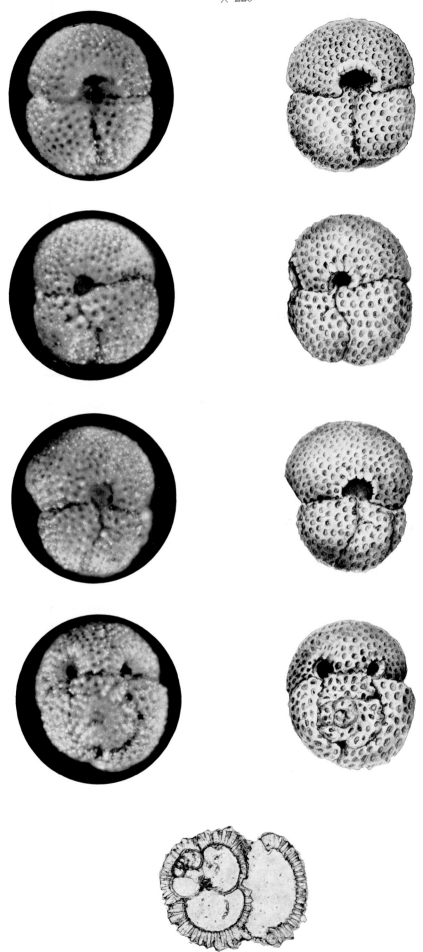

Globigerinoides extremus Bolli and Bermudez

Reference *Globigerinoides obliquus extremus* Bolli and Bermudez, 1965: Zonation based on Planktonic Foraminifera of Middle Miocene to Pliocene warm-water sediments. — Bol. Inf. Asoc. Ven. Geol., Min. Petr., 8(5):139, pl.1, figs. 10-12.

Type locality Core sample at a depth of 1,029-1,034 feet, well Cubagua No. 1, Island of Cubagua, Venezuela.

Diagnosis Test high trochospiral; equatorial periphery distinctly lobulate; axial periphery rounded. Wall distinctly perforate, surface pitted.
Chambers of the last whorl progressively more compressed in a laterally oblique manner, arranged in three to four whorls; the four chambers of the last whorl increase regularly in size, though the last one may be somewhat reduced.
Sutures slightly curved to oblique on spiral and umbilical side, depressed.
Umbilicus fairly narrow, deep.
Primary aperture interiomarginal, umbilical, a distinct arch of medium height, bordered by a rim, the last chambers show one supplementary aperture opposite the primary one.

Strat. distr. Upper part *Globorotalia acostaensis* zone into lower part *Globoquadrina altispira* zone. Questionable occurrence in upper part *G. altispira* zone.

Remarks This species differs from *Globigerinoides obliquus* Bolli in having all chambers of the last whorl compressed in a laterally oblique manner.
Locality of figured specimen is Bolli 544, Buff Bay, Jamaica, W.I.

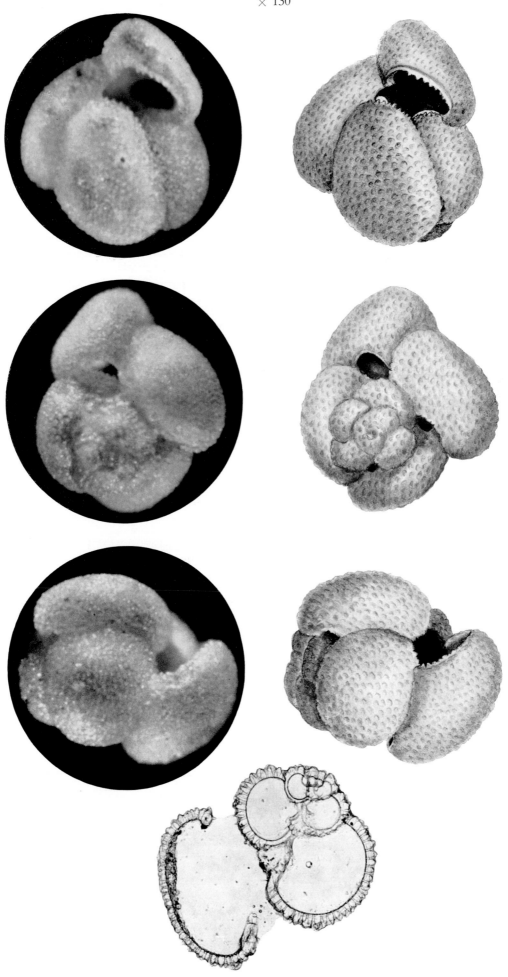

Globigerinoides fistulosus (Schubert)

Reference *Globigerina fistulosa* Schubert, 1910: Über Foraminiferen und einen Fischotolithen aus dem fossilen Globigerinenschlamm von Neu-Guinea. — Geol. Reichsanst., Verh., Wien, 1910, p.323, textfig. 1.

Type locality Siminis on Djaul (Sandwich Island), Bismarck Archipelago, Territory of New Guinea.

Diagnosis Test trochospiral, biconvex; equatorial periphery lobulate; axial periphery rounded to subangular in the last chamber.
Wall coarsely perforate, surface pitted.
Chambers spherical, except the last one or two, which are elongate, sack-like, becoming pointed, forming narrow single or multiple digitate extensions; arranged in about three whorls; the three and a half to four chambers of the last whorl increase rapidly in size. Sutures on spiral side curved, depressed; on umbilical side almost radial, depressed. Umbilicus fairly wide.
Primary aperture interiomarginal, umbilical, a wide, asymmetrical opening with a rim or lip, the secondary sutural apertures are situated over sutures of earlier chambers.

Strat. distr. Lower part of *Globoquadrina altispira* zone to Recent.

Remarks Locality of figured specimen is A 3344 (Bermudez collection of Recent material).

Globigerinoides fistulosus
$\times$ 50

293

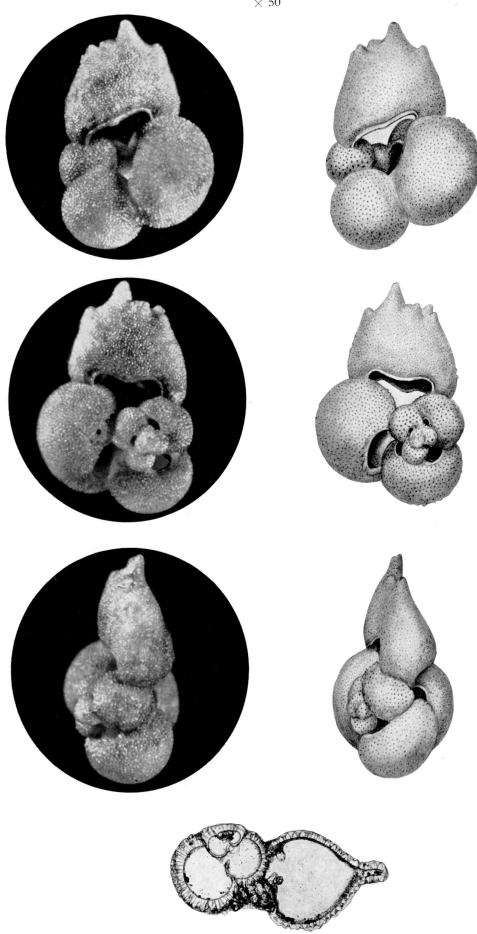

Globigerinoides immaturus Leroy

Reference *Globigerinoides sacculiferus* (Brady) var. *immatura* Leroy, 1939: Some small foraminifera, ostracoda and otoliths from the Neogene ("Miocene") of the Rokan-Tapanoeli area, Central Sumatra. — Natuurk. Tijdschr. Nederl.-Indië, 99 (6):263, pl. 3, figs. 19-21.

Type locality Locality Ho-286 A, 2.4 km N. 30° E from kampong Loeboekrambahan, Tapoeng Kiri area, Rokan-Tapanoeli Region, Central Sumatra, Indonesia.

Diagnosis Test trochospiral, unequally biconvex; equatorial periphery lobulate; axial periphery broadly rounded.
Wall distinctly perforate, surface pitted.
Chambers spherical, arranged in about three and a half whorls; the three chambers of the last whorl increase moderately in size.
Sutures on spiral side slightly curved, depressed; on umbilical side radial, depressed.
Umbilicus fairly narrow.
Primary aperture interiomarginal, umbilical, with a low to medium arch bordered by a rim; the last few chambers show one secondary sutural aperture opposite the primary aperture.

Strat. distr. Base *Globigerinoides trilobus* zone to Recent.

Remarks See remarks *Globigerinoides trilobus* (Reuss).
Locality of figured specimen is Atlantic Sta. 3490, lat. 23°11'N., long. 81°55'E.

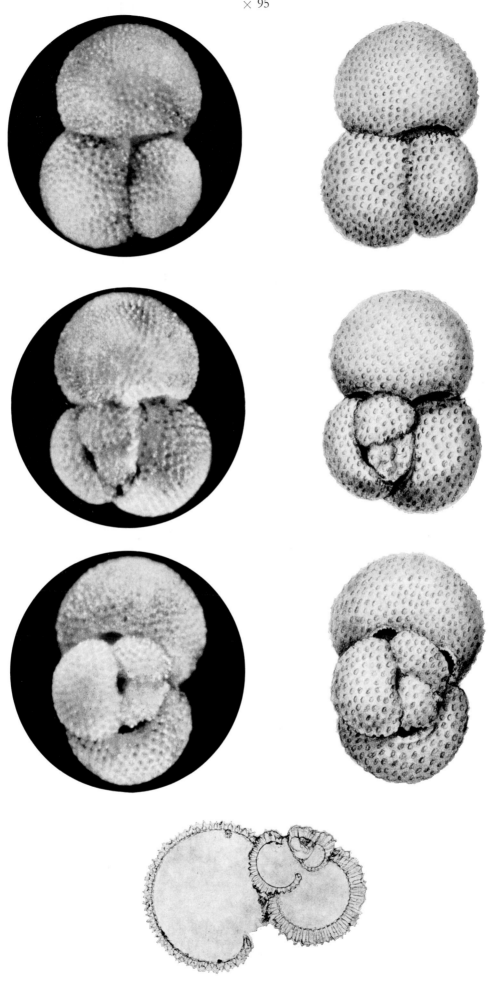

Globigerinoides obliquus BOLLI

Reference   *Globigerinoides obliqua* BOLLI, 1957: Planktonic Foraminifera from the Oligocene-Miocene Cipero and Lengua Formations of Trinidad, B.W.I. — United States National Museum Bulletin, 215:113, pl. 25, figs. 9a-10c.

Type locality   A ditch on the east side of Cunjal Road, about 150 feet from its junction with the Realize Road, about 2½ miles south-southeast of Lengua Settlement, southern Trinidad, W.I.

Diagnosis   Test trochospiral, unequally biconvex; equatorial periphery lobulate, axial periphery rounded.
Wall distinctly perforate, surface pitted.
Chambers spherical, except the ultimate one, which is compressed in a lateral oblique manner, arranged in about three and a half whorls; the three to four chambers of the last whorl increase moderately in size.
Sutures on spiral side slightly curved, depressed; on umbilical side radial, depressed.
Umbilicus narrow.
Primary aperture interiomarginal, umbilical, with a distinct, often fairly high arch bordered by a rim, the last few chambers show one secondary sutural aperture opposite the primary aperture.

Strat. distr.   Upper part *Globigerinoides trilobus* zone to upper part *Globorotalia margaritae* zone. Possibly ranging up to the *Globorotalia tosaensis* zone. Questionable occurrence in the middle part of the *Globigerinoides trilobus* zone.

Remarks   See remarks *Globigerinoides trilobus* (REUSS).
Locality of figured specimen is K.T.O.'s well W.C.C.-1,312', Trinidad, W.I.

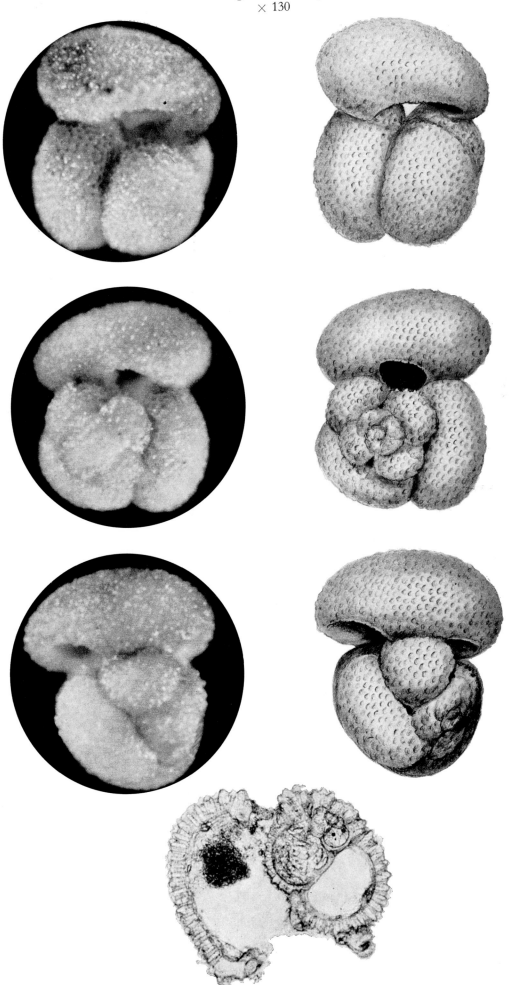

Globigerinoides primordius Blow and Banner

Reference   *Globigerinoides quadrilobatus* (D'Orbigny) *primordius* Blow and Banner, 1962: Fundamentals of Mid-Tertiary stratigraphical correlation. – Cambridge University Press: 115, pl. IX, figs. Dd-Ff.

Type locality   Holotype from the type locality of Bolli's *Globorotalia kugleri* zone, Trinidad, W.I.

Diagnosis   Test trochospiral, unequally biconvex; equatorial periphery distinctly lobulate; axial periphery rounded.
Wall perforate, surface pitted.
Chambers inflated, subglobular, arranged in two to three whorls with four chambers in the last whorl, increasing fairly rapidly in size.
Sutures on spiral and umbilical side are radial to subradial, depressed.
Umbilicus narrow.
Primary aperture interiomarginal, umbilical, a low to medium arch, bordered by a faint rim; only one secondary sutural aperture is present.

Strat. distr.   Base *Globorotalia kugleri* zone into lower part *Globigerinoides trilobus* zone. Questionable occurrence in uppermost part *Globigerina ampliapertura* zone and middle part *G. trilobus* zone.

Remarks   See remarks *Globigerina praebulloides* Blow.
Locality of figured specimen is DB 288, Trinidad, W.I.

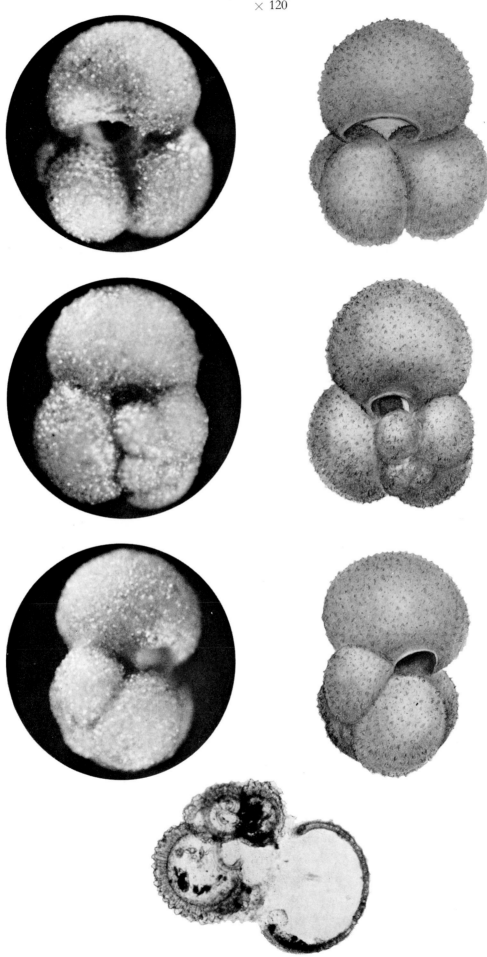

Globigerinoides ruber (D'Orbigny)

Reference *Globigerina rubra* D'Orbigny, 1839: Foraminifères. In: De la Sagra, Histoire physique, politique et naturelle de l'Ile de Cuba. – 8: 82, pl. 4, figs. 12–14.

Type locality Not designated. Localities given: Cuba, Jamaica, Guadeloupe and Martinique.

Diagnosis Test trochospiral; equatorial periphery distinctly lobulate, axial periphery broadly rounded.
Wall perforate, surface pitted.
Chambers highly inflated, spherical, arranged in three and a half to four whorls; the three chambers of the last whorl increase slowly in size and are distinctly separated.
Sutures on spiral and umbilical side subradial to radial, depressed.
Umbilicus narrow.
Primary aperture interiomarginal, umbilical, with a medium arched opening, bordered by a rim, with secondary sutural apertures situated opposite sutures of earlier chambers.

Strat. distr. From lower part *Globorotalia margaritae* zone to Recent.

Remarks See *Globigerinoides subquadratus* Brönnimann.
Locality of figured specimen is W.H.B. 187, Jamaica.

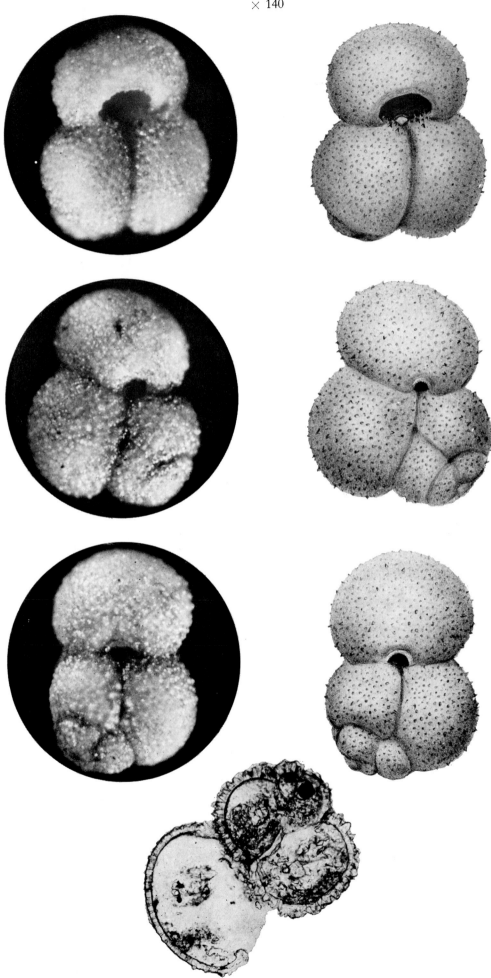

Globigerinoides sacculiferus (Brady)

Reference   *Globigerina sacculifera* Brady, 1877.: Supplementary note on the foraminifera of the Chalk (?) of the New Britain Group. — Geol. Mag., n.s., decade 2,4 (12):535 (figures in Brady, Rep. Voy. Challenger, Zool., 9, pl. 80, figs. 15-16, 1882).

Type locality   Not designated.

Diagnosis   Test trochospiral, biconvex; equatorial periphery lobulate; axial periphery rounded to subangular in the last chamber.
Wall distinctly perforate, surface pitted.
Chambers spherical, except in the last one, which is elongate, sack-like, arranged in about three and a half whorls; the three to four chambers of the last whorl increase moderately in size, the last one, however, may be rather small.
Sutures on spiral side slightly curved, depressed; on umbilical side radial, depressed.
Umbilicus fairly narrow.
Primary aperture interiomarginal, umbilical, a distinct arch bordered by a rim, the last few chambers show one secondary sutural aperture opposite the primary aperture.

Strat. distr.   Upper part *Globigerinoides trilobus* zone to Recent. Questionable occurrence in the middle part of the *Globigerinoides trilobus* zone.

Remarks   See remarks *Globigerinoides trilobus* (Reuss).
Locality of figured specimen is Bn 270, Italy.

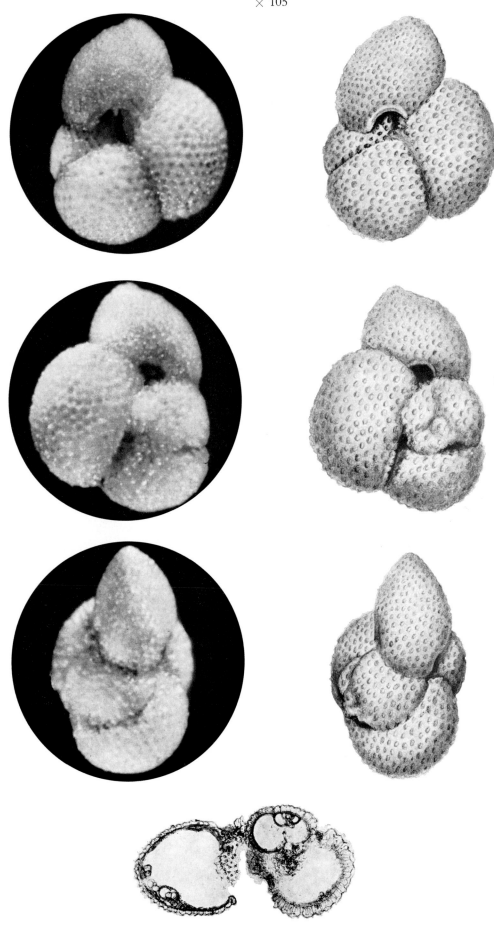

Globigerinoides sicanus De Stefani

Reference *Globigerinoides sicanus* De Stefani, 1950: Su alcune manifestazioni di idrocarburi in provincia di Palermo e descrizione di Foraminifera nuovi. – Plinia, Ital., 3 (4): 9, fig. 6 on pl. 13 of Cushman and Stainforth, 1945 (C.L.F.R. Special Pub. 14)

Type locality Sample Rz. 108, Trinidad Leaseholds Cataloque No. 21743, situated on the "Cipero nose", Cipero Marl Formation.

Diagnosis Test trochospiral; equatorial periphery moderately to slightly lobulate; axial periphery broadly rounded.
Wall distinctly perforate, surface pitted.
Chambers spherical, arranged in three and a half to four whorls; the three chambers of the last whorl increase rapidly in size, the final chamber embraces from 15 to 35 percent of the penultimate and earlier chambers.
Sutures on spiral side slightly curved to radial, slightly depressed; on umbilical side radial, moderately to slightly depressed.
Umbilicus narrow.
Aperture of the last chamber slit-like, often elongate, two to four in number, situated in the suture between the last and earlier chambers. Secondary apertures present in the sutures between the penultimate and the earlier chambers.

Strat. distr. Upper part *Globigerinatella insueta* zone and lowermost part *Globorotalia peripheroronda* zone. Questionable occurrence in middle part *G. insueta* zone.

Remarks *Globigerinoides sicanus* De Stefani is most probably developed from *Globigerinoides trilobus* (Reuss) and regarded as the ancestor of *Praeorbulina glomerosa* (Blow), *P. transitoria* (Blow) and the genus *Orbulina* (see Blow, 1956). *Globigerinoides bisphericus* Blow is a junior synonym.
Locality of figured specimen is No. 12, L.E.B., Cipero Formation, Trinidad, W.I.

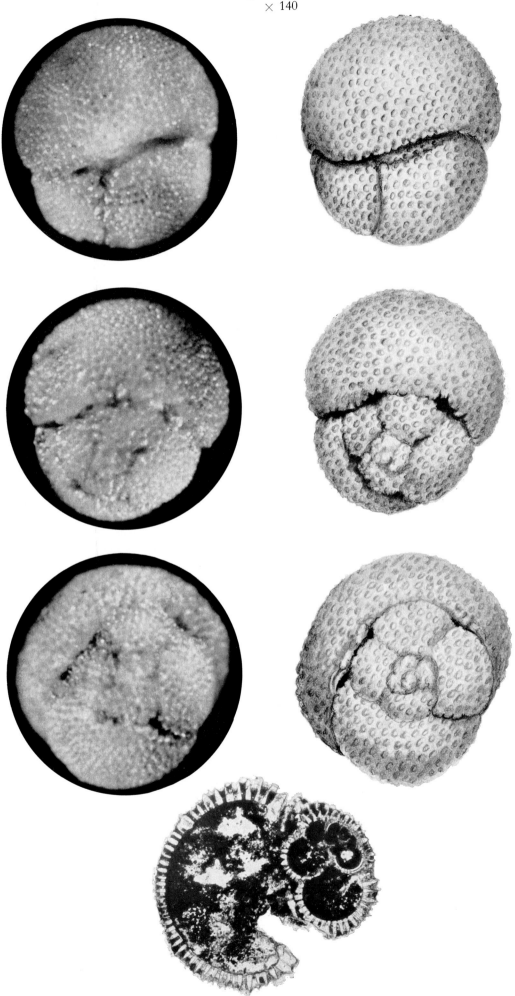

Globigerinoides subquadratus BRÖNNIMANN

Reference *Globigerinoides subquadrata* BRÖNNIMANN, 1954: Appendix: Descriptions of new species in TODD, CLOUD, LOW and SCHMIDT. – Probable occurrence of Oligocene in Saipan. – Amer. Journ. Sci., 252 (II): 680, pl. 1, figs. 8a-c.

Type locality Holotype from loc. C – 85, 1.6 miles northeast of the seaward tip of the southwest point of Saipan, Mariana Islands.

Diagnosis Test trochospiral; equatorial periphery lobulate; axial periphery broadly rounded.
Wall distinctly perforate, surface pitted.
Chambers spherical, arranged in three and a half to four whorls; the three chambers of the last whorl increase moderately in size, are slightly compressed.
Sutures on spiral side slightly curved, depressed; on umbilical side radial, depressed.
Umbilicus fairly narrow.
Primary aperture interiomarginal, umbilical, with a medium to high arch, bordered by a rim, the last few chambers show two distinct secondary sutural apertures.situated over sutures of earlier chambers.

Strat. distr. Middle part *Globigerinoides trilobus* zone to top of *G. subquadratus* zone.

Remarks *Globigerinoides subquadratus* BRÖNNIMANN differs from *Globigerinoides ruber* (D'ORBIGNY) in being more compressed and subquadratic.
Locality of figured specimen is T.W. 623, Trinidad, W.I.

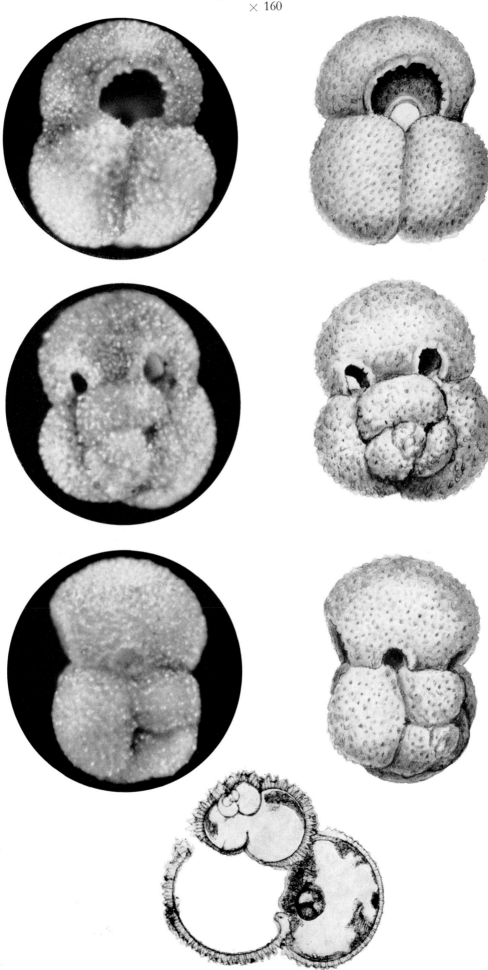

Globigerinoides trilobus (REUSS)

Reference   *Globigerina triloba* REUSS, 1850: Neue Foraminiferen aus den Schichten des Öster-
reichiches Tertiärbeckens.
Denkschr. Akad. Wiss. Wien, Math.-Nat. Classe, 1:374, pl. 47, figs. 11a-d.

Type locality   Wieliczka, Galizien, Poland.

Diagnosis   Test trochospiral, unequally biconvex; equatorial periphery lobulate; axial periphery
broadly rounded.
Wall distinctly perforate, surface pitted.
Chambers spherical, arranged in about three and a half whorls; the three chambers of
the last whorl increase rapidly in size.
Sutures on spiral side slightly curved, depressed; on umbilical side radial, depressed.
Umbilicus fairly narrow.
Primary aperture interiomarginal, umbilical, a low to medium arch, bordered by a rim,
the last few chambers show one secondary sutural aperture opposite the primary aperture.

Strat. distr.   Base *Globigerinoides trilobus* zone to Recent.

Remarks   This species is closely related to several others. *Globigerinoides trilobus* (REUSS) differs
from *G. immaturus* LEROY in having a final chamber that is larger than all earlier chambers
combined. *G. sacculiferus* (BRADY) differs from *G. immaturus* in having a terminal, elongate, sack-like chamber. *G. altiaperturus* BOLLI is distinguished from *G. trilobus* in having
a high arched primary aperture.
G. obliquus BOLLI is characterized by having the last chamber compressed in a lateral,
oblique manner.
All these species differ from *Globigerinoides ruber* (D'ORBIGNY), *G. subquadratus* BRÖNNI-
MANN and *G. diminutus* BOLLI in the position of the primary interiomarginal, umbilical
aperture and the secondary, sutural apertures.
Locality of figured specimen is Carenero-14, Venezuela.

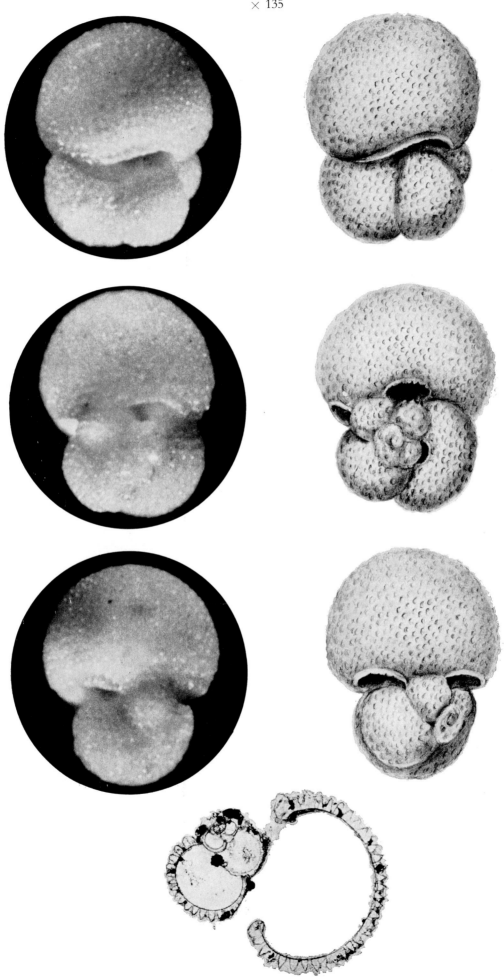

Globoquadrina altispira (CUSHMAN and JARVIS)

Reference *Globigerina altispira* CUSHMAN and JARVIS, 1936: Three new Foraminifera from the Miocene Bowden marl of Jamaica. — Contributions from the Cushman Laboratory for Foraminiferal Research, 12 (1):5, pl. 1, figs. 13a-c.

Type locality Milestone No. 71, east of Port Antonio, Jamaica.

Diagnosis Test medium to high trochospiral; equatorial periphery distinctly lobulate; axial periphery broadly rounded, with dorso-peripheral shoulders.
Wall distinctly perforate, surface finely pitted and may be hispid.
Chambers of early part spherical, those of last whorl strongly compressed laterally, arranged in three and a half to four whorls; the four to five chambers of the last whorl increase moderately in size.
Sutures on spiral side slightly curved to radial, depressed; on umbilical side radial, depressed.
Umbilicus wide to fairly wide, deep.
Aperture interiomarginal, umbilical, with high arch, covered above by an elongate, tooth-like flap.

Strat. distr. Upper part *Globigerinoides trilobus* zone to top of *Globoquadrina altispira* zone. Questionable occurrence in lower part of *G. trilobus* zone.

Remarks Locality of figured specimen is the type locality of BOLLI's *Globorotalia barisanensis* zone.

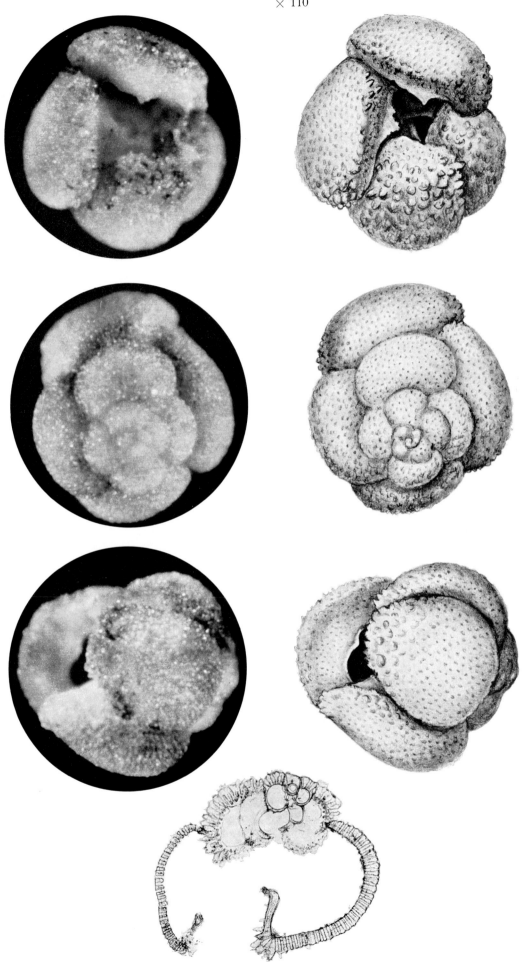

Globoquadrina dehiscens (CHAPMAN, PARR and COLLINS)

Reference *Globorotalia dehiscens* CHAPMAN, PARR and COLLINS, 1934: Tertiary foraminifera of Victoria, Australia. — The Balcombian deposits of Port Phillip; Part III — Linn. Soc. London, Journ. Zool., 38 (262):569, pl. II, figs. 36a-c.

Type locality Kackeraboite Creek, Port Phillip area, Victoria, Australia.

Diagnosis Test low trochospiral, spiral side flat, umbilical side strongly convex; equatorial periphery slightly lobulate, subquadrate; axial periphery broadly rounded, with dorso-peripheral shoulders.

Wall distinctly perforate, surface pitted, rugose near shoulders of the chambers.

Chambers of early part spherical, those of last whorl strongly compressed laterally with subrounded to angular shoulders, arranged in about four whorls; the four chambers of the last whorl increase rapidly in size and height.

Sutures on spiral side slightly curved to radial, depressed; on umbilical side radial, depressed.

Umbilicus fairly narrow to fairly wide, deep.

Aperture interiomarginal, umbilical, with a low to medium arch, covered by an elongate, tooth-like flap.

Strat. distr. Upper part of *Globigerinoides trilobus* zone into lower part of *Globorotalia acostaensis* zone. Questionable occurrence in upper part of *G. acostaensis* zone to top of *Globoquadrina altispira* zone.

Remarks Locality of figured specimen is PM 1324, Trinidad, W.I.

Globoquadrina dehiscens
× 125

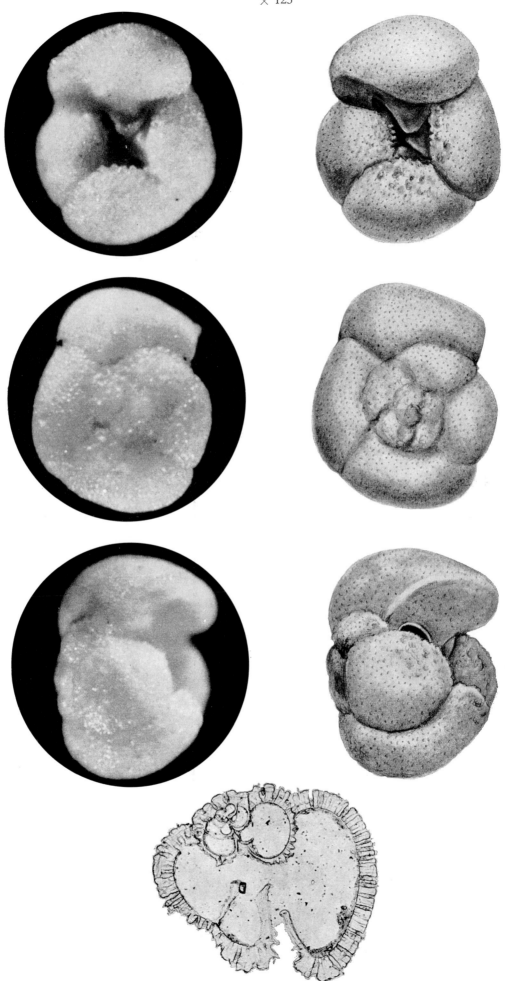

Globorotalia acostaensis BLOW

Reference *Globorotalia acostaensis* BLOW, 1959: Age, correlation and biostratigraphy of the Upper Tocuyo (San Lorenzo) and Pozón Formations, eastern Falcón, Venezuela. — Bulletins of American Paleontology, 39 (178):208, pl.17, figs. 106a-c.

Type locality Sample RM 19791, auger line near Pozón, eastern Falcón, Venezuela.

Diagnosis Test low trochospiral; equatorial periphery lobulate; axial periphery rounded.
Wall distinctly perforate, rather thick, surface pitted.
Chambers ovate to subspherical, arranged in about three whorls, with five to six chambers in the last whorl. Often the last chamber is much reduced in size and occasionally somewhat displaced towards the umbilical side.
Sutures on spiral side radial to slightly curved, depressed; on umbilical side radial, depressed.
Umbilicus narrow, deep.
Aperture interiomarginal, extraumbilical-umbilical, a low arch with usually a distinctive lip.

Strat. distr. Base *Globorotalia acostaensis* zone to top *G. tosaensis* zone. Questionable occurrence in *G. truncatulinoides* zone.

Remarks This species differs from *Globorotalia mayeri* CUSHMAN and ELLISOR in being less lobulate, in having more radial sutures, a less high arched aperture and a more distinctive apertural lip.
It differs from *Globorotalia dutertrei* (D'ORBIGNY) in having less globular and less well-separated chambers.
Locality of figured specimen is well Bodjonegoro-1, 502 m., Java, Indonesia.

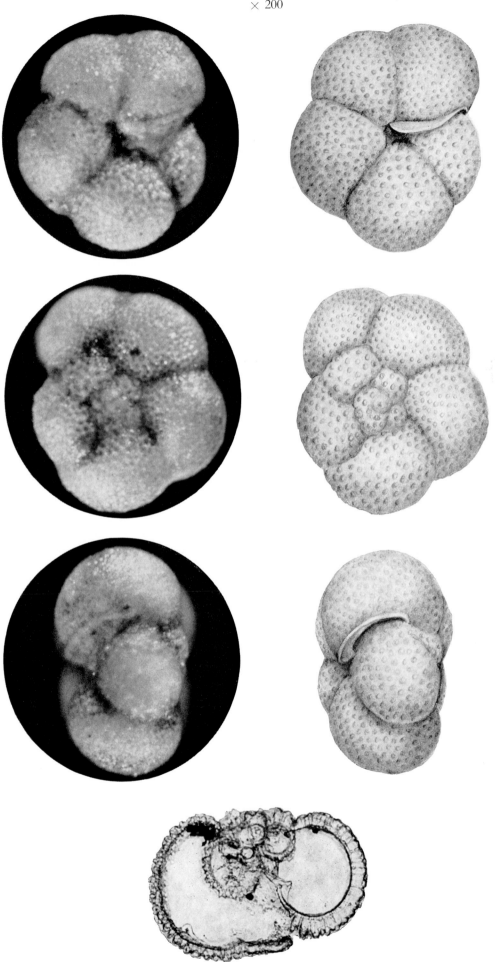

Globorotalia archeomenardii Bolli

Reference *Globorotalia archeomenardii* Bolli, 1957: Planktonic Foraminifera from the Oligocene-Miocene Cipero and Lengua Formations of Trinidad, B.W.I. — United States National Museum Bulletin, 215:119, pl. 28, figs. 11a-c.

Type locality Type locality of the *Globorotalia fohsi barisanensis* zone, Trinidad, sample Bo 202 (TTOC 193125).

Diagnosis Test low trochospiral, biconvex, compressed; equatorial periphery slightly lobulate; axial periphery acute with a rather distinct pseudo-keel.
Wall very finely perforate, surface smooth.
Chambers strongly compressed, arranged in about three whorls; the four to five chambers of the last whorl increase fairly rapidly in size.
Sutures on spiral side strongly curved, slightly depressed to flush; on umbilical side radial to slightly curved, depressed.
Umbilicus small, fairly shallow.
Aperture interiomarginal, extraumbilical-umbilical, with a low slit, bordered by a lip.

Strat. distr. Upper part of *Globigerinatella insueta* zone to top of *Globorotalia peripheroronda* zone.

Remarks *Globorotalia archeomenardii* is distinguished from *G. menardii* and *G. praemenardii* by having a more convex spiral side, in being less lobulate and smaller.
Locality of figured specimen is the type locality of the *G. peripheroronda* zone, Trinidad, W.I.

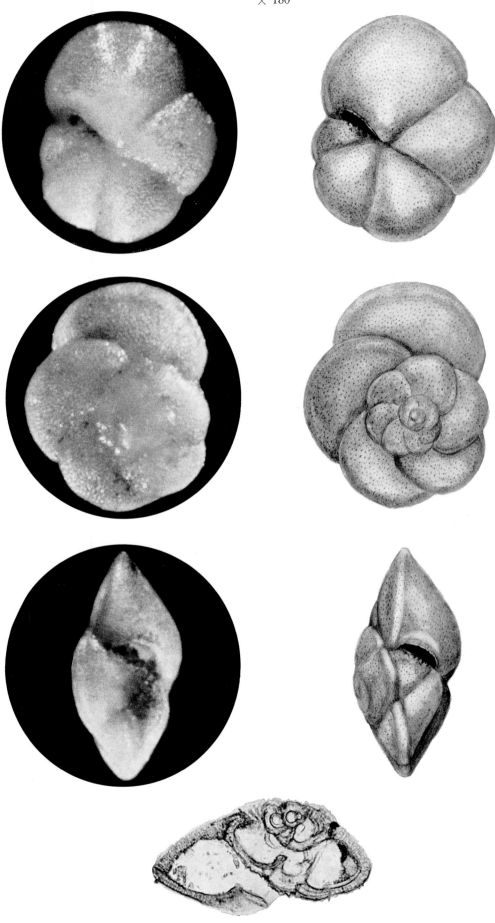

Globorotalia crassaformis (GALLOWAY and WISSLER)

Reference *Globigerina crassaformis* GALLOWAY and WISSLER, 1927: Pleistocene foraminifera from the Lomita Quarry, Palos Verdes Hills, California. — Jour. Pal., 1:41, pl.7, fig. 12.

Type locality Lomita Quarry, 2 miles south of Lomita, California, U.S.A.

Diagnosis Test low trochospiral, spiral side almost flat, inner whorl occasionally slightly raised, umbilical side strongly convex; equatorial periphery lobulate, axial periphery subacute to subrounded.

Wall finely perforate, surface of early chambers slightly rugose on umbilical as well as spiral side.

Chambers compressed, arranged in about four whorls; the four to four and a half chambers of the last whorl increase rapidly in size.

Sutures on spiral side distinctly curved, depressed; on umbilical side almost radial to slightly sigmoidal, depressed.

Umbilicus fairly narrow to fairly wide, deep.

Aperture interiomarginal, extraumbilical-umbilical, a long slit, bordered above by a lip.

Strat. distr. Upper part *Globorotalia margaritae* zone to Recent.

Remarks *Globorotalia crassaformis* is most probably the ancestor of the *Globorotalia tosaensis/truncatulinoides* group.

Globorotalia crassula CUSHMAN and STEWART is a junior synonym.

Locality of figured specimen is well Bodjonegoro - 1, 292m., Java, Indonesia.

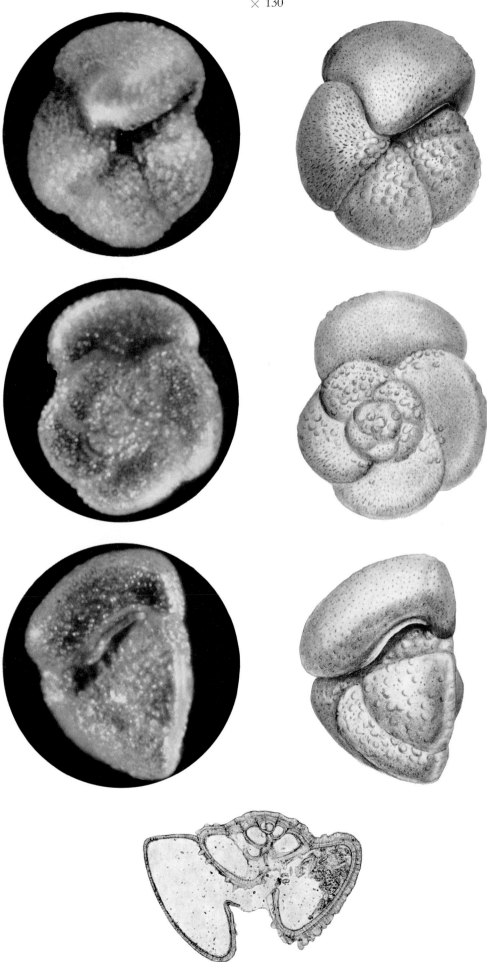

Globorotalia dutertrei (D'Orbigny)

Reference *Globigerina dutertrei* D'Orbigny, 1839: Foraminifères. In: de la Sagra, Histoire physique, politique et naturelle de l'Ile de Cuba. — 8:84, pl.4, figs. 19-21.

Type locality Not designated. Localities given: recent marine sands of Cuba, Martinique and Guadaloupe.

Diagnosis Test low to medium high trochospiral; equatorial periphery moderately lobulate; axial periphery rounded.
Wall moderately to coarsely perforate, fairly thick; surface pitted.
Chambers inflated, arranged in about three whorls; the five to six chambers of the last whorl increase hardly in size.
Sutures on spiral side almost radial to slightly curved, depressed; on umbilical side almost radial, depressed.
Umbilicus fairly wide, deep.
Aperture interiomarginal, extraumbilical-umbilical, a fairly high arch bordered by a narrow rim.

Strat. distr. Base *Globorotalia dutertrei* zone to Recent.

Remarks *Globorotalia eggeri* (Rhumbler), *G. eggeri multiloba* (Romeo) and *G. dubia* (Egger) are probably junior synonyms of, or anyhow closely related to *Globorotalia dutertrei*.
Locality of figured specimen is 524 "Downwind", BG 137, lat. 9° 53' S., long. 110° 41' W., Pacific Ocean.

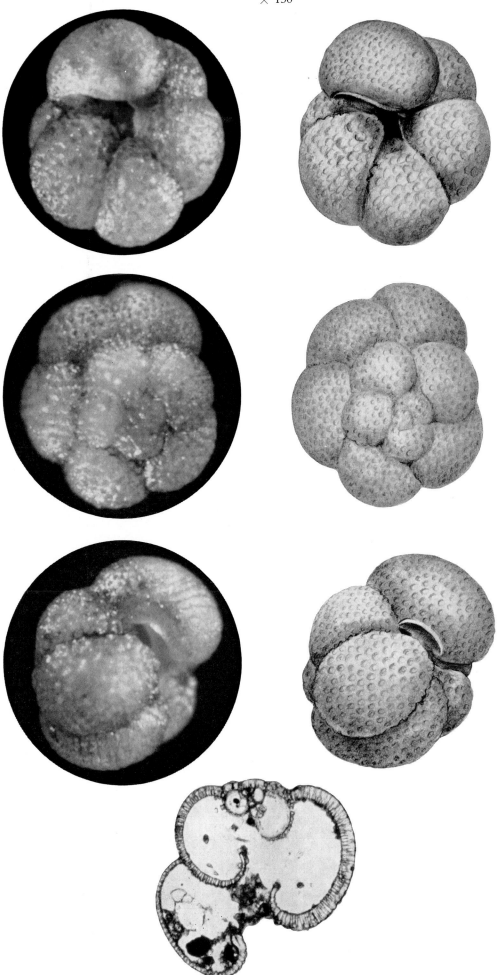

Globorotalia fohsi CUSHMAN and ELLISOR

Reference   *Globorotalia fohsi* CUSHMAN and ELLISOR, 1939: New species of Foraminifera from the Oligocene and Miocene. – Contributions from the Cushman Laboratory for Foraminiferal Research, 15 (1): 12, pl. 2, figs. 6a-c.

Type locality   Core sample at a depth of 9,612 feet from Humble Oil and Refining Company's No. 1 Ellender, Terrebonne Parish, Louisiana, U.S.A.

Diagnosis   Test very low trochospiral, unequally biconvex, compressed; equatorial periphery slightly lobulate; axial periphery acute with a distinct keel.
Wall distinctly perforate, surface of early chambers rugose, that of later chambers smooth. Chambers compressed, arranged in about three whorls; the six to nine chambers of the last whorl increase regularly in size.
Sutures on spiral side curved, flush to raised; on umbilical side slightly curved to radial, depressed.
Umbilicus fairly narrow, fairly deep.
Aperture interiomarginal, extraumbilical-umbilical, with a low arch, bordered by a lip.

Strat. distr.   Ranging throughout the *Globorotalia fohsi* zone.

Remarks   We follow mainly the views of BLOW and BANNER (1966) concerning the *Globorotalia fohsi* group. However, there appears to be no reason to redesignate *Globorotalia lobata* BERMUDEZ as *G. praefohsi* BLOW and BANNER.
The following names are used in this manual: *Globorotalia peripheroronda* BLOW and BANNER (=*G. fohsi barisanensis* LEROY, sensu BOLLI), *Globorotalia peripheroacuta* BLOW and BANNER (=*G. fohsi fohsi* CUSHMAN and ELLISOR, sensu BOLLI), *Globorotalia lobata* BERMUDEZ (=*G. praefohsi* BLOW and BANNER p.p.), *Globorotalia fohsi* CUSHMAN and ELLISOR (=*G. fohsi robusta* BOLLI).
Locality of figured specimen is the type locality of BOLLI's *Globorotalia fohsi robusta* zone, Trinidad, W.I.

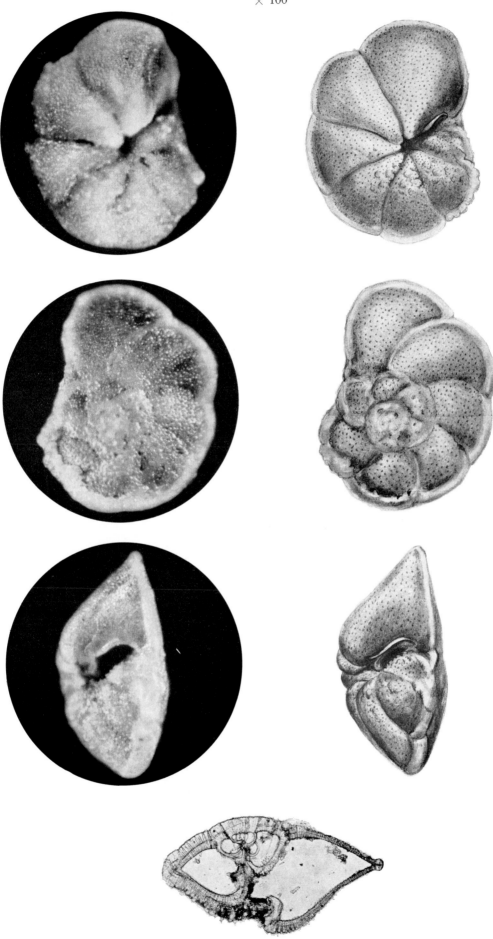

Globorotalia kugleri Bolli

Reference *Globorotalia kugleri* Bolli, 1957: Planktonic Foraminifera from the Oligocene-Miocene Cipero and Lengua Formations of Trinidad, B.W.I. — United States National Museum Bulletin, 215:118, pl. 28, figs. 5a-6.

Type locality Along the south bank of the San Fernando bypass road, approximately 240 feet northeast of the north end of the road bridge across the Siparia railway line, near San Fernando, western Trinidad.

Diagnosis Test very low trochospiral, unequally biconvex; equatorial periphery slightly lobulate; axial periphery rounded or with a tendency to become subangular.
Wall distinctly perforate, surface of early chambers may be very slightly rugose on umbilical side, that of later chambers smooth.
Chambers ovate, arranged in about three whorls; the six to eight chambers of the last whorl increase slowly in size.
Sutures on spiral side curved, depressed; on umbilical side radial, distinctly depressed. Umbilicus fairly narrow.
Aperture interiomarginal, extraumbilical-umbilical, with a distinct arch, bordered by a lip.

Strat. distr. Base of *Globorotalia kugleri* zone into lowermost part of *Globigerinoides trilobus* zone. Questionable occurrence in rest of lower part of *G. trilobus* zone.

Remarks This fairly small species is distinguished from *Globorotalia peripheroronda* Blow and Banner by the constant length and greater number of chambers in the last whorl.
Locality of figured specimen is the type locality of the *Globorotalia kugleri* zone, Trinidad, W.I.

$\times$ 220

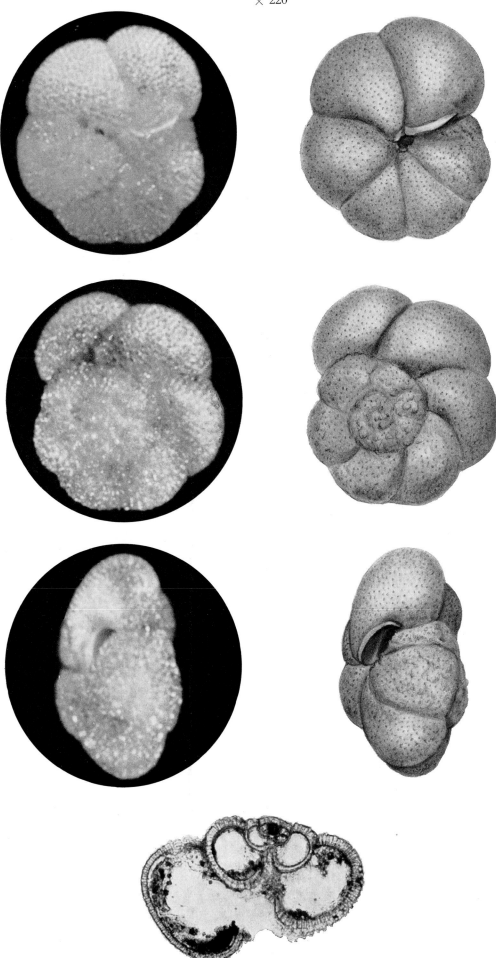

Globorotalia lenguaensis BOLLI

Reference *Globorotalia lenguaensis* BOLLI, 1957: Planktonic Foraminifera from the Oligocene-Miocene Cipero and Lengua Formations of Trinidad, B.W.I. — United States National Museum Bulletin, 215:120, pl.29, figs. 5a-c.

Type locality Type locality of the *Globorotalia menardii* zone, sample KR 23425 (TTOC 178890).

Diagnosis Test low trochospiral, about equally biconvex; equatorial periphery almost circular, axial periphery subangular to angular.
Wall finely perforate, surface smooth.
Chambers strongly compressed, the six to seven chambers of the last whorl increase moderately in size.
Sutures on spiral side curved, occasionally slightly depressed; on umbilical side radial to slightly sigmoidal, depressed.
Umbilicus almost closed, shallow.
Aperture interiomarginal, extraumbilical-umbilical, a low arch, bordered by a sometimes pronounced lip.

Strat. distr. Base of *Globorotalia siakensis* zone into lower part of *Globorotalia acostaensis* zone. Questionable occurrence in upper part of *Globigerinoides subquadratus* zone and middle part of

Remarks Locality of figured specimen is the above mentioned type locality.

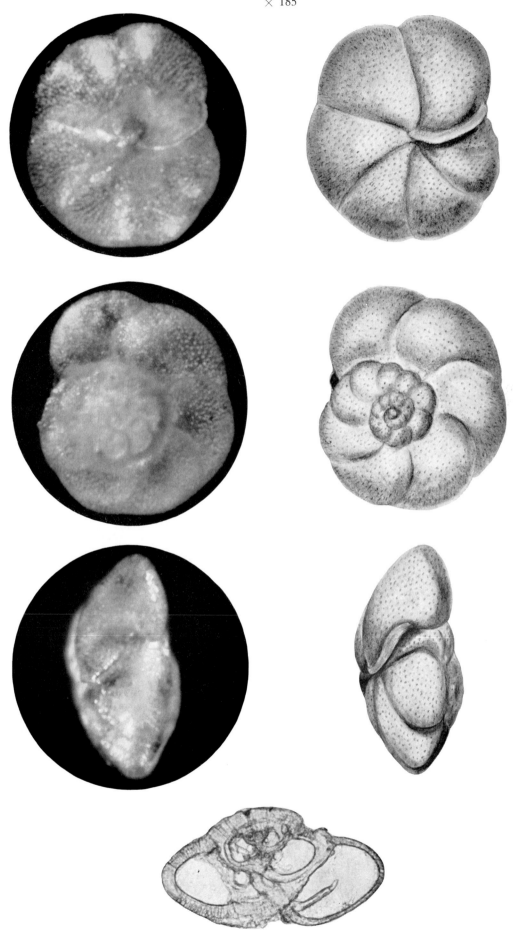

Globorotalia lobata BERMUDEZ

Reference — *Globorotalia lobata* BERMUDEZ, 1949: Tertiary smaller foraminifera of the Dominican Republic. — Cushman Laboratory for Foraminiferal Research, Special Publ., 25:286, pl. 22. figs. 15-17.

Type locality — Core sample at a depth of 74-84 feet, from Bravo well no. 5, Yaguate area, Trujillo Province. Dominican Republic.

Diagnosis — Test very low trochospiral, biconvex, compressed; equatorial periphery of the last chambers distinctly lobulate, axial periphery of the earlier part subacute to slightly rounded, in the later portion acute, giving the impression of a keel.
Wall distinctly perforate, surface of early chambers may be slightly rugose, that of later chambers smooth.
Chambers compressed, arranged in about three whorls; the six to eight chambers of the last whorl increase rather rapidly in size.
Sutures on spiral side strongly curved, raised, on umbilical side slightly curved to radial, depressed.
Umbilicus fairly narrow, shallow.
Aperture interiomarginal, extraumbilical-umbilical, a low arch, bordered by a distinct lip.

Strat. distr. — Uppermost part *Globorotalia peripheroacuta* zone into lowermost part *G. fohsi* zone.

Remarks — See remarks *Globorotalia fohsi* CUSHMAN and ELLISOR.
Locality of figured specimen is TW 629, Trinidad, W.I.

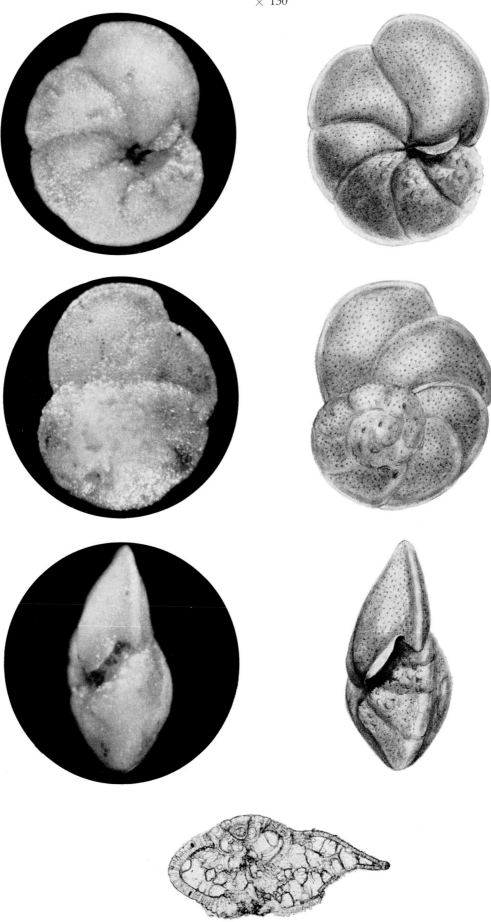

Globorotalia margaritae BOLLI and BERMUDEZ

Reference *Globorotalia margaritae* BOLLI and BERMUDEZ, 1965: Zonation based on Planktonic Foraminifera of Middle Miocene to Pliocene warm-water sediments. — Bol. Inf. Asoc. Ven. Geol., Min. Petr., 8 (5):139, pl.1, figs. 16-18.

Type locality Cut in road Porlamar to Boca del Rio, east of Espinal and immediately west of junction where highway to Punta de Piedras branches off, Isla Margarita, Venezuela.

Diagnosis Shape of test low trochospiral, spiral side convex, umbilical side less convex to almost flat; equatorial periphery slightly lobulate; axial periphery acute with a thin keel.
Wall finely perforate, surface of early chambers slightly rugose, later ones smooth.
Chambers strongly compressed, arranged in about three whorls; the five to six chambers of the last whorl increase rapidly in size, the last chamber constituting a considerable part on the umbilical side of the surface of the test.
Sutures on spiral side strongly curved, flush to slightly raised; on umbilical side almost radial, slightly depressed.
Umbilicus very narrow to closed, shallow.
Aperture interiomarginal, extraumbilical-umbilical, a low slit bordered by a lip.

Strat. distr. Ranging throughout the *Globorotalia margaritae* zone.

Remarks Locality of figured specimen is Bolli 544, Buff Bay, Jamaica, W.I.

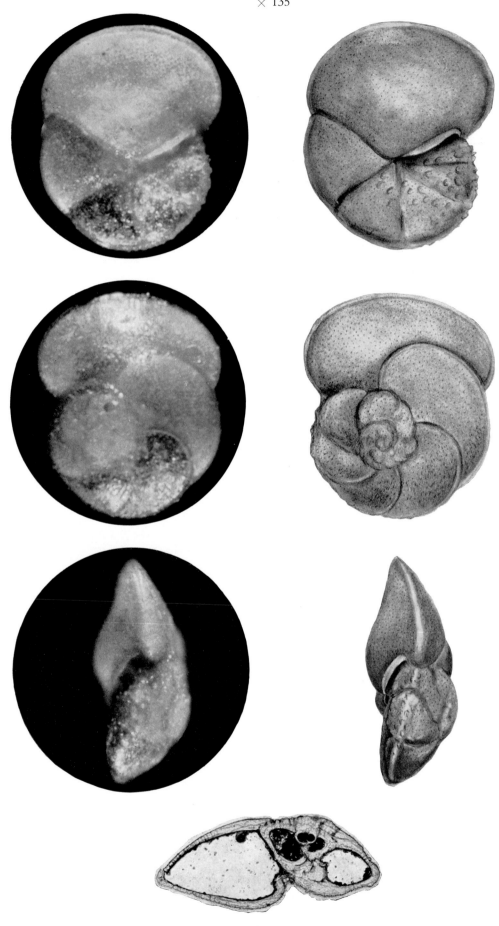

Globorotalia mayeri CUSHMAN and ELLISOR

Reference   *Globorotalia mayeri* CUSHMAN and ELLISOR, 1939: New species of foraminifera from the Oligocene and Miocene. — Contributions from the Cushman Laboratory for Foraminiferal Research, 15 (1):11, pl. 2, figs. 4a-c.

Type locality   Core sample at a depth of 9,612 feet, from Humble Oil and Refining Company's No. 1 Ellender, Terrebonne Parish, Louisiana, U.S.A.

Diagnosis   Test very low trochospiral, inflated; equatorial periphery lobulate; axial periphery broadly rounded.
Wall rather coarsely perforate, surface smooth.
Chambers inflated, subglobular, arranged in about three whorls; the five to six chambers of the last whorl increase regularly in size.
Sutures on spiral side slightly to moderately curved, depressed; on umbilical side radial, depressed.
Umbilicus fairly wide and deep.
Aperture interiomarginal, extraumbilical-umbilical, with a large, high arch, bordered by a lip or rim.

Strat. distr.   Upper part *Globorotalia peripheroronda* zone into lower part *Globigerinoides subquadratus* zone. Questionable occurrence in the upper part of the *G. subquadratus* zone.

Remarks   This species differs from *Globorotalia peripheroronda* BLOW and BANNER in having more inflated chambers and a higher arched aperture; it differs from *G. siakensis* LEROY in having curved dorsal sutures and less separated chambers.
Locality of figured specimen is Hermitage, Trinidad, W.I.

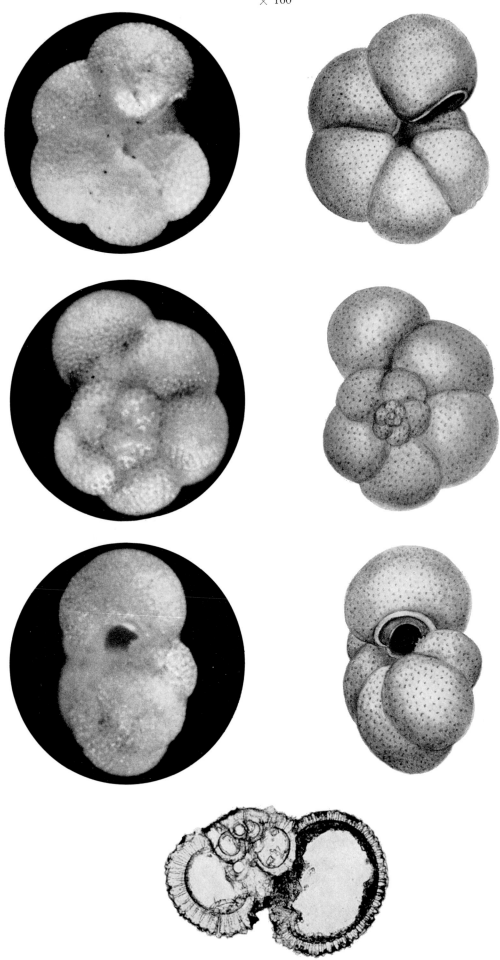

Globorotalia menardii (D' Orbigny)

Reference *Rotalia menardii* D'Orbigny, 1826; Tableau méthodique de la classe des Céphalopodes. — Annales des Sciences Naturelles, Paris, France, Sér. 1, tome 7:273, modèle No. 10.

Type locality Adriatic Sea, near Rimini, Italy.

Diagnosis Test very low trochospiral, biconvex, compressed; equatorial periphery lobulate, axial periphery acute with a pronounced keel.
Wall finely perforate, surface of early chambers slightly rugose near shoulders, later ones smooth.
Chambers strongly compressed, arranged in about three whorls, the five to seven chambers of the last whorl increase regularly in size.
Sutures on spiral side strongly curved, raised, on umbilical side radial to slightly curved, depressed.
Umbilicus fairly wide, shallow.
Aperture interiomarginal, extraumbilical-umbilical, a low slit, bordered by a distinct lip.

Strat. distr. Upper part of *Globorotalia fohsi* zone to Recent.

Remarks This species differs from *Globorotalia praemenardii* Cushman and Stainforth in having a more pronounced keel and more distinct chambers.
Locality of figured specimen is Pacific Ocean, lat. 9°53' S., long. 110°41' W., depth 2770 m.

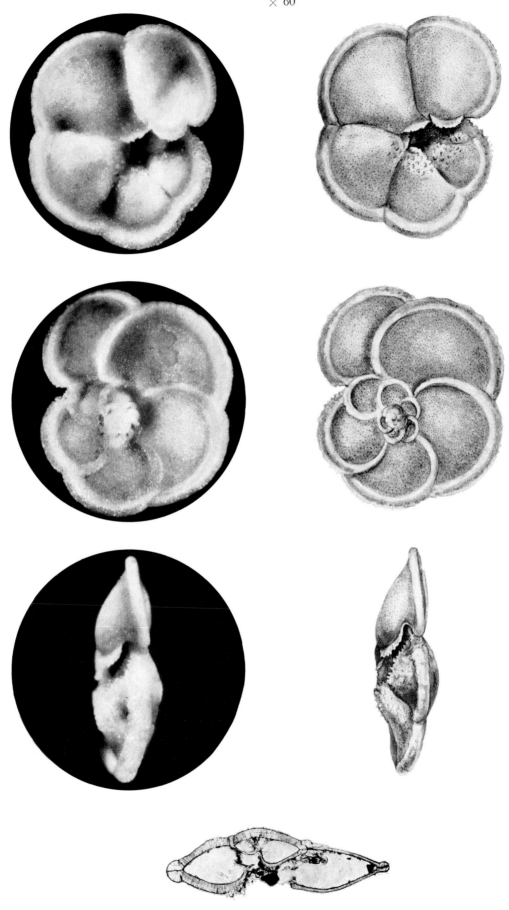

Globorotalia miocenica PALMER

Reference   *Globorotalia menardii* (D'ORBIGNY) var. *miocenica* PALMER, 1945: Notes on the Foraminifera from Bowden, Jamaica. — Bull. Amer. Paleont., 29 (115):70, pl.1, figs. 10a-c.

Type locality   K. V. Palmer Sta. 1, Port Morant, at foot of hill where road to old Capt. Baker house turns off main road to Bowden P.O. and the United Fruit Co. wharf, Bowden, Jamaica, W.I.

Diagnosis   Test very low trochospiral, spiral side flat; umbilical side strongly convex; equatorial periphery almost circular in the early chambers to very slightly lobulate in the later stage; axial periphery acute with a distinct keel.
Wall finely perforate, surface of umbilical area of early chambers slightly rugose, that of later chambers smooth.
Chambers angular, the six to seven chambers of the last whorl increase moderately in size.
Sutures on spiral side broadly curved, slightly raised; on umbilical side radial to slightly curved, slightly depressed.
Umbilicus fairly narrow, fairly deep.
Aperture interiomarginal, extraumbilical-umbilical, a fairly low slit, bordered by a rim or lip.

Strat. distr.   Base *Globorotalia margaritae* zone to top *Globoquadrina altispira* zone. Questionable occurrence in upper part *Globorotalia dutertrei* zone and in lowermost part *G. tosaensis* zone.

Remarks   The species differs from *Globorotalia pseudomiocenica* BOLLI and BERMUDEZ in having a less lobulate equatorial periphery, a somewhat wider umbilicus and in having a completely flat spiral side.
Locality of figured specimen is the Bowden Formation type locality, Jamaica, W.I.

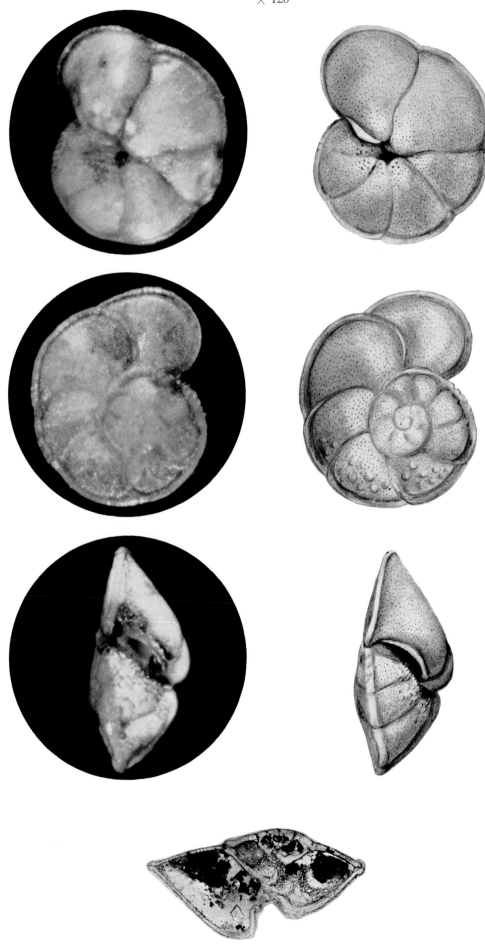

Globorotalia multicamerata CUSHMAN and JARVIS

Reference *Globorotalia menardii* (D'ORBIGNY) var. *multicamerata* CUSHMAN and JARVIS, 1930: Miocene foraminifera from Buff Bay, Jamaica. — Jour. Pal., 4 (4):367, pl.34, figs. 8a-c.

Type locality Half mile east of Buff Bay, Jamaica, W.I.

Diagnosis Test very low trochospiral, biconvex, compressed; equatorial periphery weakly lobulate; axial periphery acute with a very pronounced, broad keel.
Wall finely perforate, surface generally smooth, umbilical area of early chambers may be slightly rugose.
Chambers strongly compressed, arranged in about three whorls; the seven to eight chambers of the last whorl increase slowly or hardly in size.
Sutures on spiral side of early chambers strongly curved, of later chambers less so, raised; on umbilical side radial to slightly curved, depressed.
Umbilicus fairly narrow, shallow.
Aperture interiomarginal, extraumbilical-umbilical, a low slit, bordered by a distinct lip.

Strat. distr. Base *Globorotalia margaritae* zone into lower part *Globorotalia tosaensis* zone. Questionable occurrence in uppermost part *Globorotalia dutertrei* zone.

Remarks This species differs from *Globorotalia menardii* (D'ORBIGNY) in having more chambers in the last whorl, a somewhat narrower keel and in the heavy development of the peripheral keel.
Locality of figured specimen is the above mentioned type locality.

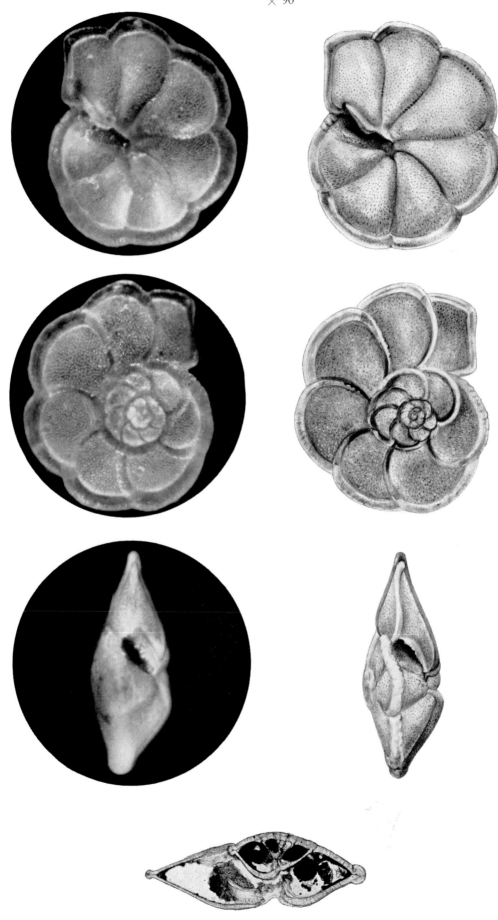

Globorotalia nana BOLLI

Reference *Globorotalia opima nana* BOLLI, 1959: Planktonic Foraminifera from the Oligocene-Miocene Cipero and Lengua Formations of Trinidad, B.W.I. — United States National Museum Bulletin, 215:118, pl. 28, figs. 3a-c.

Type locality A surface sample from the northern part of the Cipero type section, 20-240 feet southeast of the southernmost iron rail which serves as the fixed point for that type section, along the Cipero coast and south of San Fernando, western Trinidad, W.I.

Diagnosis Test very low trochospiral; equatorial periphery slightly lobulate; axial periphery rounded. Wall rather coarsely perforated, surface pitted.
Chambers spherical, arranged in about three whorls; the four to five chambers of the last whorl increase fairly rapidly in size.
Sutures on spiral side almost radial, depressed; on umbilical side radial, depressed.
Umbilicus narrow, deep.
Aperture interiomarginal, extraumbilical-umbilical, with a low arch, bordered by a thick rim or lip.

Strat. distr. Base of *Globigerina ampliapertura* zone into lowermost part of *Globorotalia kugleri* zone. Questionable occurrence in the Upper Eocene.

Remarks This species is distinguished from *Globorotalia opima* by being considerably smaller. Locality of figured specimen is DB 293, Trinidad, W.I.

Globorotalia nana
× 180

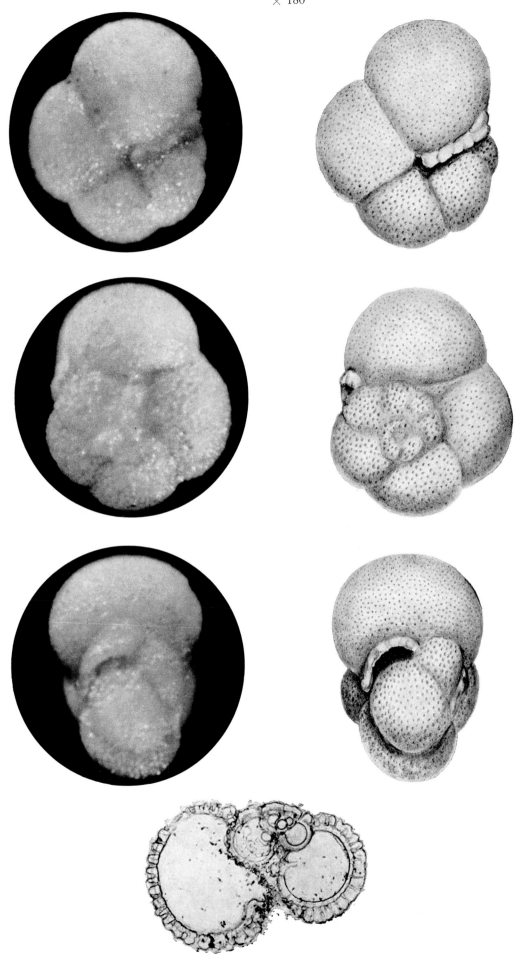

Globorotalia obesa BOLLI

Reference
Globorotalia obesa BOLLI, 1957: Planktonic Foraminifera from the Oligocene-Miocene Cipero and Lengua Formations of Trinidad, B.W.I. — United States National Museum Bulletin, 215:119, pl. 29, figs. 2a-3.

Type locality
Surface sample JS 16 (TTOC 193261), 850-1400 feet southwest of the southernmost iron rail which serves as the fixed point for the Cipero type section, Trinidad, W.I.

Diagnosis
Test very low trochospiral; equatorial periphery distinctly lobulate; axial periphery rounded.
Wall coarsely perforate, surface pitted, that of early chambers may be slightly rugose.
Chambers strongly inflated, spherical, arranged in two and a half to three whorls; generally four chambers are present in the last whorl; these increase very rapidly in size.
Sutures on spiral side and umbilical side radial, depressed.
Umbilicus fairly wide, deep.
Aperture interiomarginal, extraumbilical-umbilical, with a medium to high arch, bordered by a slight rim or lip.

Strat. distr.
Lower part of *Globigerinoides trilobus* zone to upper part of *Globorotalia margaritae* zone. Questionable occurrence from upper part *G. margaritae* zone to Recent.

Remarks
Locality of figured specimen is PM 1324, Trinidad, W.I.

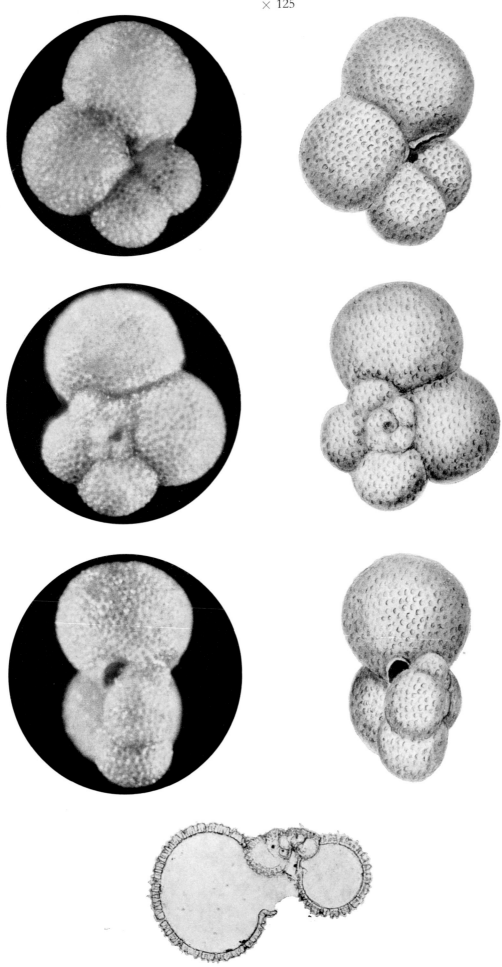

Globorotalia opima BOLLI

Reference *Globorotalia opima* BOLLI, 1957: Planktonic Foraminifera from the Oligocene-Miocene Cipero and Lengua Formations of Trinidad, B.W.I. — United States National Museum Bulletin, 215:117, pl. 28, figs. 1a-2.

Type locality A surface sample from the northern part of the Cipero type section, 20-240 feet southeast of the southernmost iron rail which serves as the fixed point for that type section, along the Cipero coast and south of San Fernando, western Trinidad, W.I.

Diagnosis Test very low trochospiral, equatorial periphery slightly lobulate; axial periphery rounded. Wall rather coarsely perforated, surface of early chambers slightly rugose on umbilical side, that of later chambers pitted.
Chambers spherical, arranged in about three whorls; the four to five chambers of the last whorl increase fairly rapidly in size.
Sutures on spiral side almost radial, depressed; on umbilical side radial, depressed.
Umbilicus narrow, deep.
Aperture interiomarginal, extraumbilical-umbilical, with a low arch, bordered by a slight rim or lip.

Strat. distr. Lower part of *Globigerina angulisuturalis* zone.

Remarks This species differs from *Globorotalia siakensis* LEROY in having fewer chambers in the last whorl, a lower arched aperture and in being "fatter" as the name indicates.
Locality of figured specimen is the above-mentioned type locality.

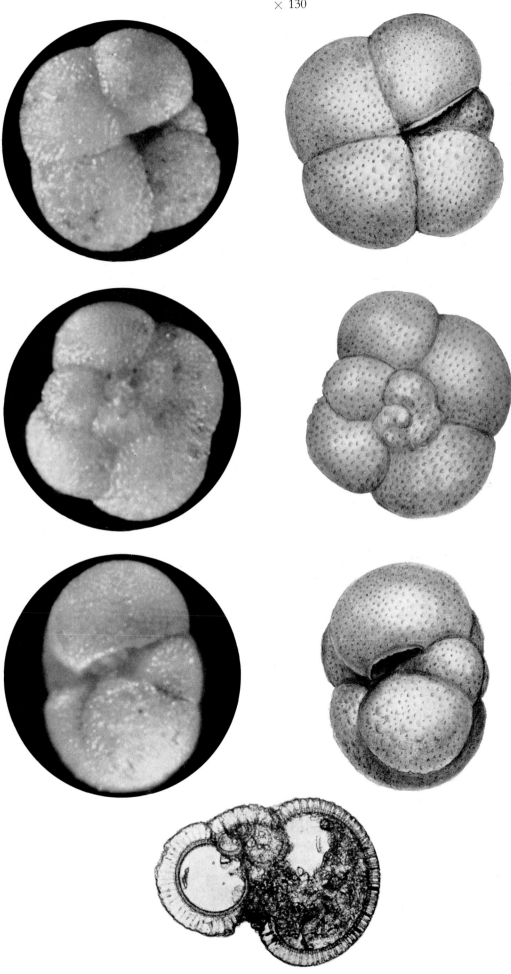

Globorotalia peripheroacuta Blow and Banner

Reference   *Globorotalia (Turborotalia) peripheroacuta* Blow and Banner, 1966: The morphology, taxonomy and biostratigraphy of *Globorotalia barisanensis* Leroy, *Globorotalia fohsi* Cushman and Ellisor, and related taxa. - Micropaleontology, 12 (3): 294. pl. 1, figs. 2a-c.

Type locality   Sample RM. 19367, El Mene-Pozón Road traverse, eastern Falcón, Venezuela, Husito Marly Clay Member, Pozón Formation.

Diagnosis   Test very low trochospiral, unequally biconvex, compressed; equatorial periphery slightly lobulate; axial periphery of the earlier part subacute to slightly rounded, in the later portion more acute.
Wall distinctly perforate, surface smooth, sometimes surface of early chambers slightly rugose.
Chambers compressed, arranged in about three whorls; the six to seven chambers of the last whorl increase regularly in size.
Sutures on spiral side strongly curved, flush to slightly depressed; on umbilical side slightly curved to radial, depressed.
Umbilicus narrow.
Aperture interiomarginal, extraumbilical-umbilical, with a distinct arch, bordered by a prominent lip.

Strat. distr.   Uppermost part of *Globorotalia peripheroronda* zone to top of *G. peripheroacuta* zone. Questionable occurrence in middle part of *G. peripheroronda* zone.

Remarks   See remarks *Globorotalia fohsi* Cushman and Ellisor.
Locality of figured specimen is the type locality of Bolli's *G. fohsi fohsi* zone, Trinidad, W.I.

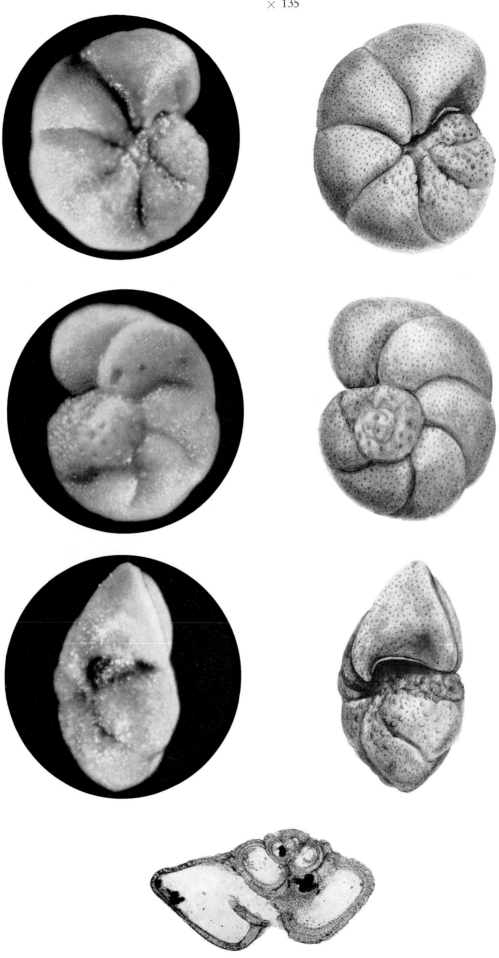

Globorotalia peripheroronda Blow and Banner

Reference *Globorotalia (Turborotalia) peripheroronda* Blow and Banner, 1966: The morphology, taxonomy and biostratigraphy of *Globorotalia barisanensis* Leroy, *Globorotalia fohsi* Cushman and Ellisor, and related taxa. – Micropaleontology, 12 (3): 294, pl. 1, figs. 1a-c.

Type locality Sample RM. 19304, El mene-Pozón Road traverse, eastern Falcón, Venezuela, Husito Marly Clay Member, Pozón Formation.

Diagnosis Test very low trochospiral, unequally biconvex; equatorial periphery slightly lobulate; axial periphery rounded or with a tendency to become subangular.
Wall distinctly perforate, surface smooth, sometimes surface of early chambers slightly rugose.
Chambers ovate, arranged in about three whorls; the usually six chambers of the last whorl increase rather slowly in size, with the exception of last chamber which may be considerably larger.
Sutures on spiral side strongly curved, slightly depressed; on umbilical side slightly curved to radial, depressed.
Umbilicus narrow.
Aperture interiomarginal, extraumbilical-umbilical, with a rather low arch, bordered by a distinct lip.

Strat. distr. Lower part *Globigerinoides trilobus* zone to top of *Globorotalia peripheroronda* zone.

Remarks See remarks *Globorotalia fohsi* Cushman and Ellisor.
Locality of figured specimen is the type locality of Bolli's *Globorotalia fohsi barisanensis* zone, Trinidad, W.I.

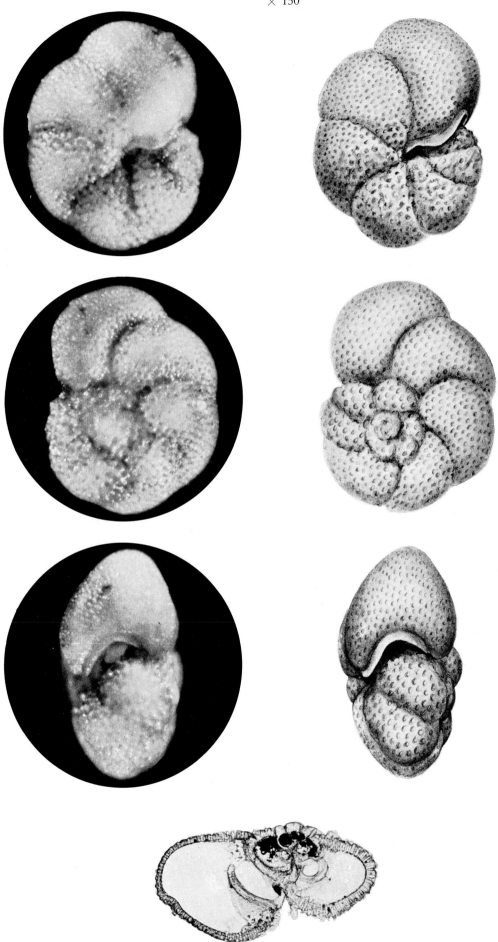

Globorotalia plesiotumida BLOW and BANNER

Reference *Globorotalia (G.) tumida* (BRADY) *plesiotumida* BLOW and BANNER, 1965: Two new taxa of the Globorotaliinae (Globigerinacea, Foraminifera) assisting determination of the Late Miocene/Middle Miocene boundary. — Nature, 207 (5004):1353, figs. 2a-c.

Type locality Core sample at a depth of 2,700 feet, well Cubagua No. 2, Island of Cubagua, Venezuela.

Diagnosis Test very low trochospiral, unequally biconvex, compressed; equatorial periphery slightly lobulate; axial periphery acute with a distinct keel.
Wall finely perforate; most of the surface is smooth, but granules are developed on the area of the wall of the first three chambers which immediately faces the aperture.
Chambers compressed, arranged in three whorls; the five to six chambers of the last whorl increase regularly in size.
Sutures on spiral side curved, later ones becoming almost sub-radial, their distal parts being strongly curved, flush to raised; on umbilical side radial to slightly curved, shallowly depressed.
Umbilicus narrow to closed, deep.
Aperture interiomarginal, extraumbilical-umbilical, a rather low arch, bordered by a thick lip.

Strat. distr. Base *Globorotalia dutertrei* zone to upper part *G. margaritae* zone. Questionable occurrence in uppermost part *G. margaritae* zone.

Remarks *Globorotalia tumida* (BRADY) differs from *Globorotalia plesiotumida* in having a larger test at the same growth stage, with a more rapid increase in whorl height, in possessing a more massive keel, thicker walls, greater rugosity of the early chambers of the ventral side, a stronger convexity of the spiral side and a higher aperture with a very broad lip.
Globorotalia pseudomiocenica differs from *G. plesiotumida* in having a smaller test at the same growth stage, a slower increase in whorl height, more uniformly enlarging chambers, a thinner keel, thinner walls and more consistently oblique sutures on the spiral side.
Locality of figured specimen is WHB 187B, Jamaica, W.I.

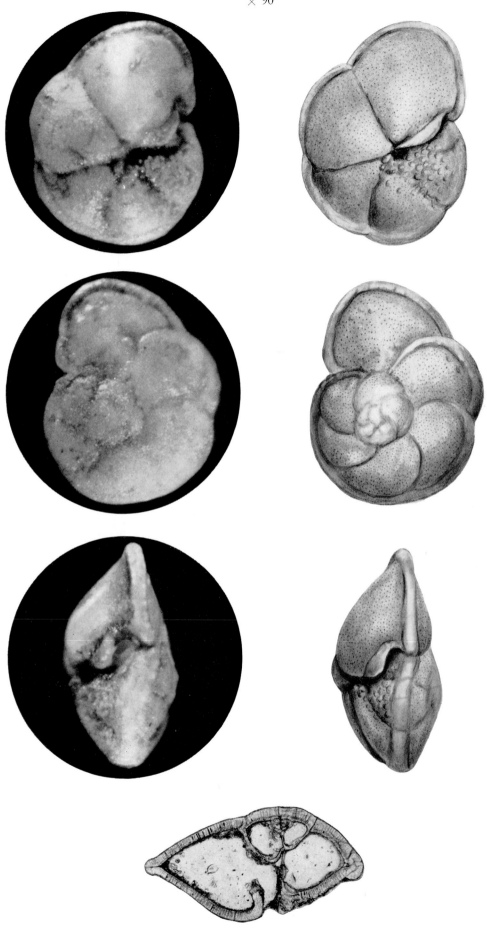

Globorotalia praemenardii CUSHMAN and STAINFORTH

Reference *Globorotalia praemenardii* CUSHMAN and STAINFORTH, 1945: The foraminifera of the Cipero marl formation of Trinidad, British West Indies. — Cushman Laboratory for Foraminiferal Research, Special Publ., 14:70, pl. 13, figs. 14a-c.

Type locality Surface sample Rz. 425, T.L.C. No. 61418, along the coast between the mouth of the Cipero River and the point at which the Trinidad Government railway turns inland, south of San Fernando, Trinidad, W.I.

Diagnosis Test very low trochospiral, biconvex, compressed; equatorial periphery moderately lobulate; axial periphery acute with a rather distinct pseudo-keel.
Wall very finely perforate, surface smooth.
Chambers strongly compressed, arranged in about three whorls; the five to six chambers of the last whorl increase fairly rapidly in size.
Sutures on spiral side strongly curved, flush to slightly raised; on umbilical side radial to slightly curved, depressed.
Umbilicus narrow, fairly shallow.
Aperture interiomarginal, extraumbilical-umbilical, with a low slit, bordered by a lip.

Strat. distr. Uppermost part of *Globorotalia peripheroronda* zone to top of *G. fohsi* zone.

Remarks This species differs from *Globorotalia archeomenardii* BOLLI in having a less convex spiral side, in being larger and more lobulate; it differs from *Globorotalia menardii* (D'ORBIGNY) in having a less pronounced keel and less depressed sutures.
Locality of figured specimen is TW 623, Trinidad, W.I.

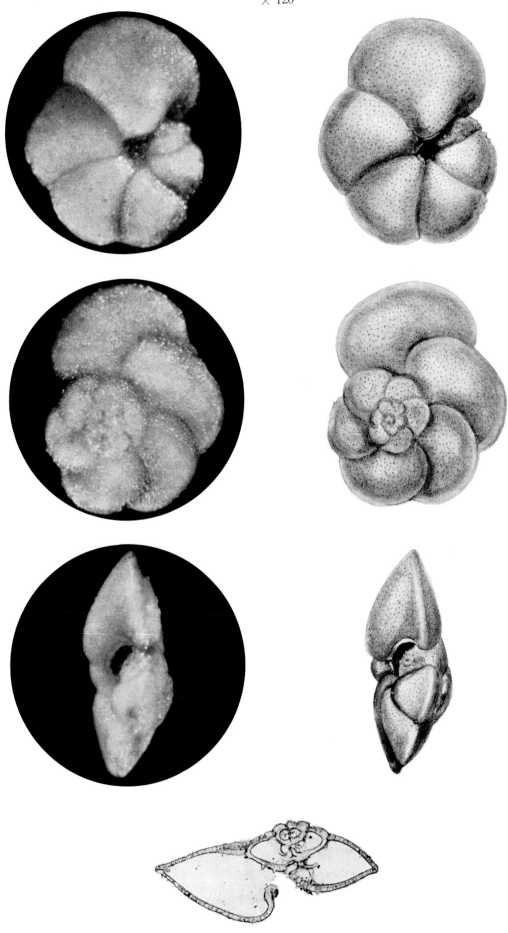

Globorotalia pseudomiocenica BOLLI and BERMUDEZ

Reference   *Globorotalia pseudomiocenica* BOLLI and BERMUDEZ, 1965: Zonation based on Planktonic Foraminifera of Middle Miocene to Pliocene warm-water sediments. — Bol. Inf. Asoc. Ven. Geol. Min. Petr., 8 (5):140, pl.1, figs. 13-15.

Type locality   Cut in road Higueroto to Chirimena, after bridge across lagoon immediately northwest of Carenero, Estado Miranda, Venezuela. *Globorotalia menardii* zone, Carenero Formation.

Diagnosis   Test very low trochospiral, spiral side flat to very slightly convex, umbilical side strongly convex; equatorial periphery slightly lobulate; axial periphery acute with a thin but distinct keel.

Wall finely perforate, surface of early chambers slightly rugose, that of later chambers smooth.

Chambers angular, somewhat compressed, arranged in about three whorls; the five to six chambers of the last whorl increase moderately in size.

Sutures on spiral side curved, slightly raised; on umbilical side radial to slightly curved, slightly depressed.

Umbilicus narrow, shallow.

Aperture interiomarginal, extraumbilical-umbilical, a low slit, bordered by a fine rim or lip.

Strat. distr.   Base *Globorotalia acostaensis* zone to lower part *G. margaritae* zone. Questionable occurrence in upper part *G. siakensis* zone, *G. menardii* zone and middle part *G. margaritae* zone.

Remarks   This species differs from *Globorotalia miocenica* PALMER in having a more lobulate equatorial periphery, a more narrow umbilicus, while the spiral side of typical *Globorotalia miocenica* is always completely flat. *Globorotalia merotumida* BLOW and BANNER is most probably a junior synonym.

Locality of figured specimen is the above mentioned type locality.

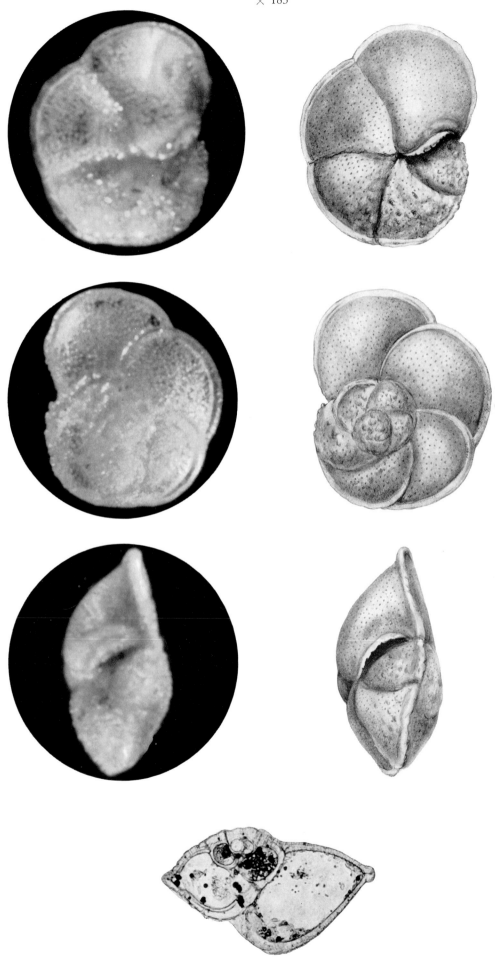

Globorotalia scitula (BRADY)

Reference   *Pulvinulina scitula* BRADY, 1882: Report on the foraminifera. In: TIZARD and MURRAY, Exploration of the Faröe Channel during the summer of 1880, in Her Majesty's hired ship "Knight Errant". — Roy. Soc. Edinburgh, Proc., II (III):716 (figures in BRADY, Rep. Voy. Challenger, Zool., pl. 103, figs. 7a-c, 1882).

Type locality   Not designated. Faröe Channel.

Diagnosis   Test low trochospiral, about equally biconvex; equatorial periphery slightly lobulate; axial periphery subangular to angular, sometimes with a pseudo-keel.

Wall finely perforate, surface of early chambers near the shoulders may be slightly rugose, that of later chambers smooth.

Chambers strongly compressed, arranged in about three whorls; the four to five chambers of the last whorl increase moderately to rapidly in size.

Sutures on spiral side strongly curved, slightly raised, on umbilical side radial to slightly curved, depressed.

Umbilicus narrow, shallow.

Aperture interiomarginal, extraumbilical-umbilical, with a low slit, bordered by a rim or lip.

Strat. distr.   Upper part *Globorotalia peripheroacuta* zone to Recent. *Globorotalia* cf. *scitula* ranges from upper part *Globigerinoides trilobus* zone to upper part of *G. peripheroacuta* zone.

Remarks   *Globorotalia* cf. *scitula* is most probably *Globorotalia scitula praescitula* BLOW. It has more tangentially elongate chambers and a more convex spiral side than *Globorotalia scitula*.

Locality of figured specimen is the type locality of BOLLI's *Globorotalia fohsi robusta* zone, Trinidad, W.I.

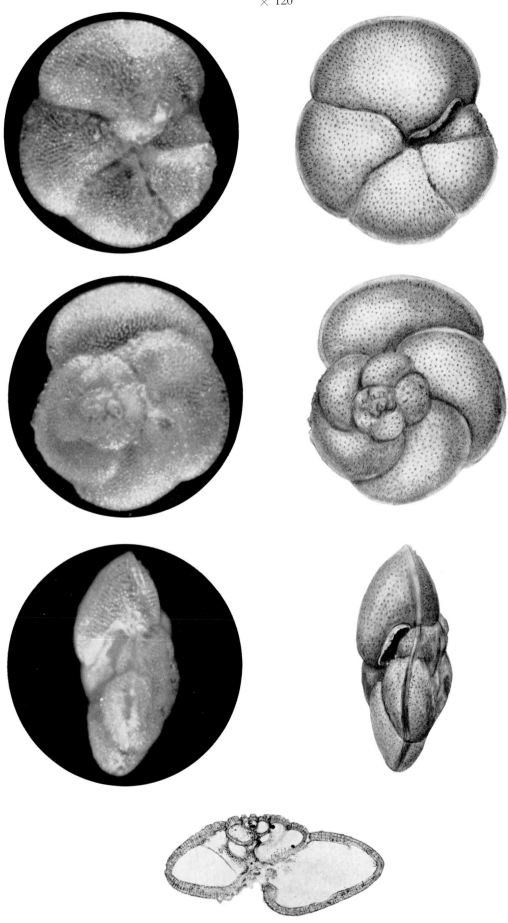

Globorotalia siakensis (LEROY)

Reference *Globigerina siakensis* LEROY, 1939: Some small foraminifera, ostracoda and otoliths from the Neogene ("Miocene") of the Rokan-Tapanoeli area, Central Sumatra. – Natuurk. Tijdschr. Nederl.-Indië, 99 (6): 262, pl. 4, figs. 20–22.

Type locality Locality Ho 528, 5 km. N. 33° E. from kampong Aliantan, Tapoeng Kiri area, Rokan-Tapanoeli region, Central Sumatra, Indonesia.

Diagnosis Test very low trochospiral, equatorial periphery lobulate; axial periphery broadly rounded.
Wall rather coarsely perforate, surface smooth.
Chambers inflated, subglobular, slightly embracing, arranged in about three whorls; the five to six chambers of the last whorl increase regularly in size.
Sutures on spiral and umbilical side radial, depressed.
Umbilicus fairly wide to fairly narrow, deep.
Aperture interiomarginal, extraumbilical-umbilical, a fairly low, elongated arched opening, bordered by a faint lip or rim.

Strat. distr. Upper part *Globigerina angulisuturalis* zone to top *Globorotalia siakensis* zone. Questionable occurrence in lower part *G. angulisuturalis* zone.

Remarks See *Globorotalia mayeri* CUSHMAN and ELLISOR.
Locality of figured specimen is DB 94, Trinidad, W.I.

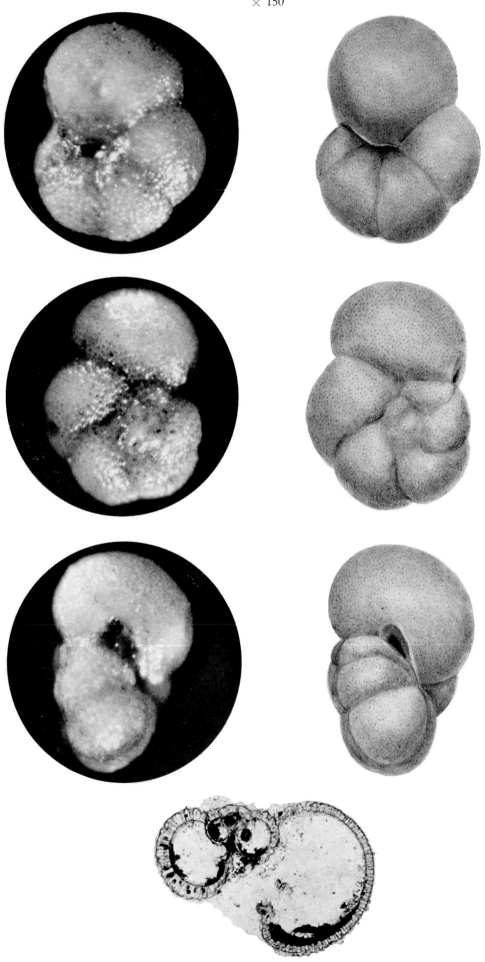

Globorotalia tosaensis TAKAYANAGI and SAITO

Reference *Globorotalia tosaensis* TAKAYANAGI and SAITO, 1962: Planktonic Foraminifera from the Nobori Formation, Shikoku, Japan. — Science Rep. Tohoku University, Second Series, Sp. Vol. 5:81, pl.28, figs. 11a-12c.

Type locality 100 meters east of Nobori, Muroto City, eastern coast of Tosa Bay, Shikoku, Japan.

Diagnosis Test very low trochospiral, spiral side almost flat, umbilical side strongly convex; equatorial periphery almost circular to slightly lobulate, axial periphery subrounded to subacute.
Wall finely perforate, surface of early chambers on umbilical side rather rugose.
Chambers subinflated, compressed laterally, giving the test a conical appearance, arranged in about three and a half whorls; the four to five chambers of the last whorl increase fairly rapidly in size.
Sutures on spiral side gently curved, somewhat indistinct, flush to slightly depressed; on umbilical side almost radial to somewhat sigmoidal, depressed.
Umbilicus fairly narrow to fairly wide, deep.
Aperture interiomarginal, extraumbilical-umbilical, a low arch bordered above by a narrow lip.

Strat. distr. Uppermost part *Globoquadrina altispira* zone and *Globorotalia tosaensis* zone. Questionable occurrence in *Globorotalia truncatulinoides* zone.

Remarks See *Globorotalia truncatulinoides* (D'ORBIGNY).
Locality of figured specimen is well Bodjonegoro — 1, 221m., Java, Indonesia.

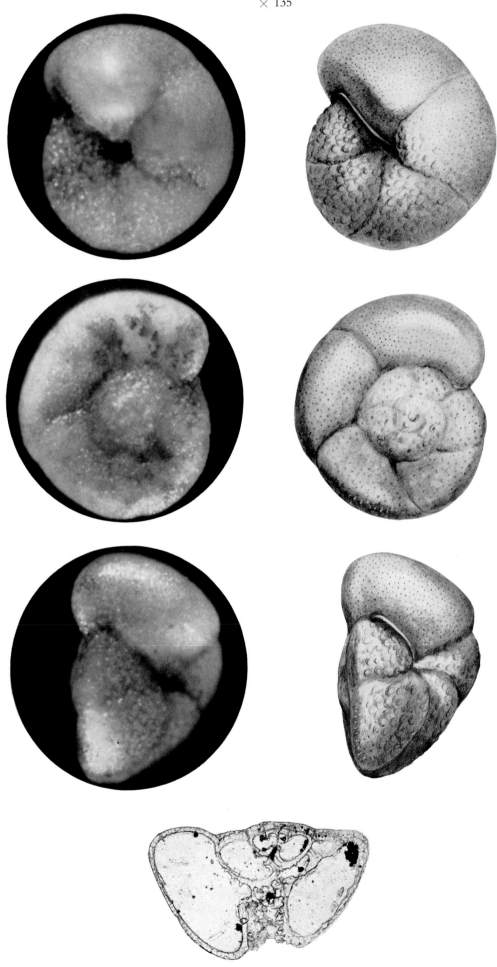

Globorotalia truncatulinoides (D'Orbigny)

Reference   *Rotalia truncatulinoides* D'Orbigny, 1839: Foraminifères. In: Barker-Webb and Berthelot, Hist. Nat. Isles Canaries. — 2 (2):132, pl.2, figs. 25-27.

Type locality   Ténériffe, Canary Islands.

Diagnosis   Test very low trochospiral, spiral side flat to slightly concave, umbilical side strongly convex; equatorial periphery almost circular to slightly lobulate, axial periphery acute with a distinct keel.
Wall finely perforate, surface of early chambers on umbilical side rugose.
Chambers compressed, giving the test a distinctly conical appearance, arranged in about three and a half whorls; the four to five chambers of the last whorl increase rapidly in size, the last chamber showing a somewhat truncate shape.
Sutures on spiral side gently curved, flush to slightly depressed; on umbilical side almost radial to somewhat sigmoidal, depressed.
Umbilicus fairly wide, deep.
Aperture interiomarginal, extraumbilical-umbilical, a low arch bordered above by a lip.

Strat. distr.   Base *Globorotalia truncatulinoides* zone to Recent.

Remarks   This species differs from *Globorotalia tosaensis* Takayanagi and Saito in having an acute axial periphery with a keel, a more distinct concave spiral side, and more pronounced, sharper shoulders.
Locality of figured specimen is 523 "Downwind", HG 74, lat. 28°29'S., long. 106°30' W., Pacific Ocean.

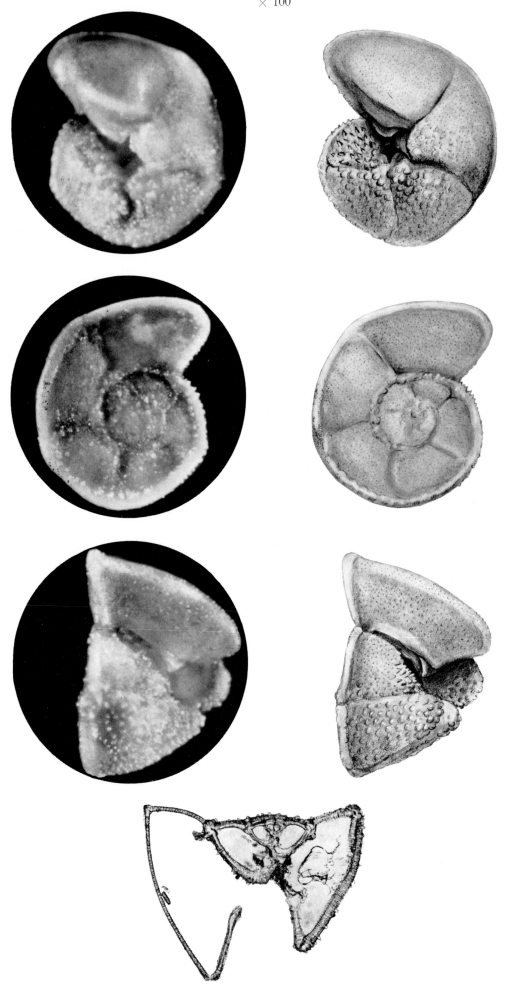

Globorotalia tumida (BRADY)

Reference *Pulvinulina menardii* (D'ORBIGNY) var. *tumida* BRADY, 1877: Supplementary note on the foraminifera of the Chalk (?) of the New Britain Group. — Geol. Mag., London, n.s., dec. 2, vol. 4, p. 535. Type figure not given, see BRADY, 1884, Rept. Voy. Challenger, Zool., vol. 9, pl.103, figs. 4-6 (as *Pulvinulina tumida* BRADY).

Type locality From a fragment of soft, white calcareous rock, found on a beach on the east side of New Ireland (Territory of New Guinea), Bismarck Archipelago. Brady's figured specimens from Challenger Sta. 224, lat. 7° 45' N., long. 144° 20' E., North Pacific Ocean, and Sta. 276, lat. 13° 28' S., long. 149° 30' W., South Pacific Ocean.

Diagnosis Test moderately low trochospiral, spiral side more convex than umbilical side, compressed; equatorial periphery subcircular to slightly lobulate in the final stage; axial periphery acute with a massive keel.
Wall finely perofarte, thick; surface smooth except for the umbilical side of the early chambers of the last whorl and the umbilical margins of the later chambers, which are heavily pustulose.
Chambers compressed, arranged in about three whorls; the six chambers of the last whorl increase regularly in size.
Sutures on spiral side initially smoothly curved, later more sharply curved, their proximal ends almost straight and nearly radial, their distal ends re-curved almost tangentially to the periphery, raised; on umbilical side almost radial to slightly sinuous, depressed.
Umbilicus fairly narrow, deep.
Aperture interiomarginal, extraumbilical-umbilical, a rather high arch, partially covered by a broad, thick lip.

Strat. distr. Base *Globorotalia margaritae* zone to Recent.

Remarks See remarks *Globorotalia plesiotumida* BLOW and BANNER. *Globorotalia flexuosa* (KOCH) is most probably an environmental manifestation of *G. tumida* (BRADY), having the same stratigraphic distribution.
Locality of figured specimen is 524 "Downwind", BG 137, lat. 9° 53' S., long. 110° 41' W., Pacific Ocean.

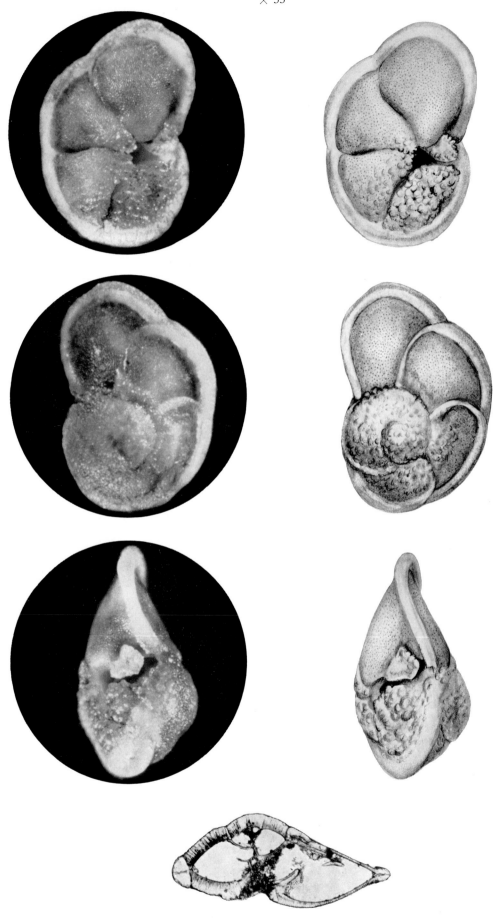

Hastigerina aequilateralis (Brady)

Reference	*Globigerina aequilateralis* Brady, 1879: Notes on some of the reticularian Rhizopoda of the "Challenger" Expedition; II — Additions to the knowledge of porcellanous and hyaline types. — Quart. Jour. Micr. Sci., n.s., 19:71 (figures in Brady, Rep. Voy. Challenger, Zool., 9, pl.80, figs. 18-21).
Type locality	Not designated. One locality given: Sta. 224, lat. 7° 45' N., long. 144° 20' E., in the North Pacific Ocean, in 1850 fathoms.
Diagnosis	Test planispiral in adult stage, trochospiral in early stage, evolute; equatorial periphery distinctly lobulate, axial periphery rounded. Wall perforate, surface finely pitted, may be slightly hispid. Chambers globular to subglobular, arranged in about three and a half whorls; the five to six chambers of the last whorl increase rapidly in size. Sutures radial, depressed. Aperture in later stage interiomarginal, equatorial, a broad arch, bordered by a faint rim.
Strat. distr.	Lower part of *Globorotalia siakensis* zone to Recent.
Remarks	When found in Miocene samples this species is usually recorded as *Hastigerina* or *Globigerinella* cf. *aequilateralis*, because it is slightly more involute than Brady's Recent types. Locality of figured specimen is from the above mentioned Challenger Station 224.

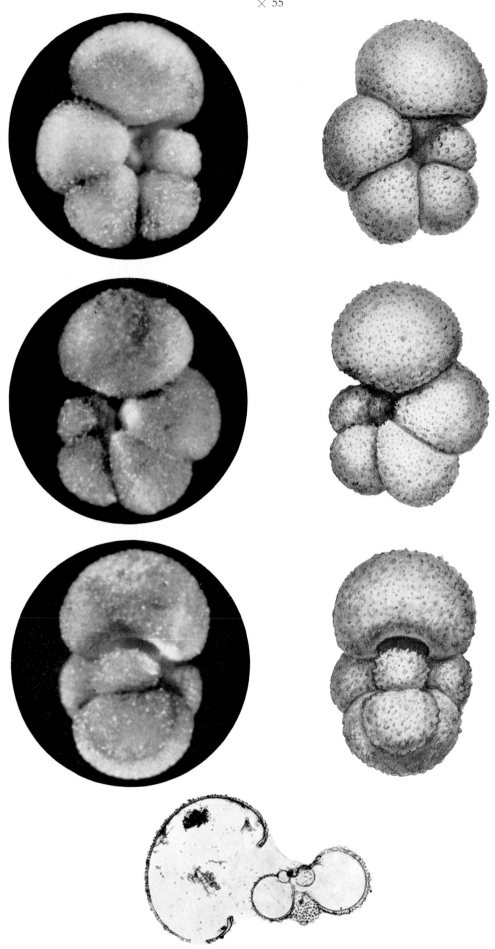

Hastigerinella bermudezi BOLLI

Reference   *Hastigerinella bermudezi* BOLLI, 1957: Planktonic Foraminifera from the Oligocene-Miocene Cipero and Lengua Formations of Trinidad, B.W.I. — United States National Museum Bulletin, 215:112, pl. 25, figs. 1a-c.

Type locality   Type locality of BOLLI's *Globorotalia fohsi barisanensis* zone, Trinidad, sample Bo. 202 (TTOC 193125).

Diagnosis   Test very low trochospiral; equatorial periphery very strongly lobulate.
Wall perforate, surface finely pitted.
Chambers in early stage spherical to ovate, the ultimate ones becoming club-shaped, arranged in two and a half to three whorls; the four to five chambers of the last whorl increase rapidly in size.
Sutures on spiral side slightly curved to radial, depressed; on umbilical side radial, depressed.
Umbilicus fairly wide, shallow.
Aperture interiomarginal, extraumbilical-umbilical, with an elongate arch or slit extending up the apertural face, bordered by a lip.

Strat. distr.   Ranging from base to upper part of *Globorotalia peripheroronda* zone. Questionable occurrence in uppermost part of *Globigerinatella insueta* zone and uppermost part of *G. peripheroronda* zone.

Remarks   Locality of figured specimen is the type locality of BOLLI's *Globorotalia fohsi barisanensis* zone, Trinidad, W.I.

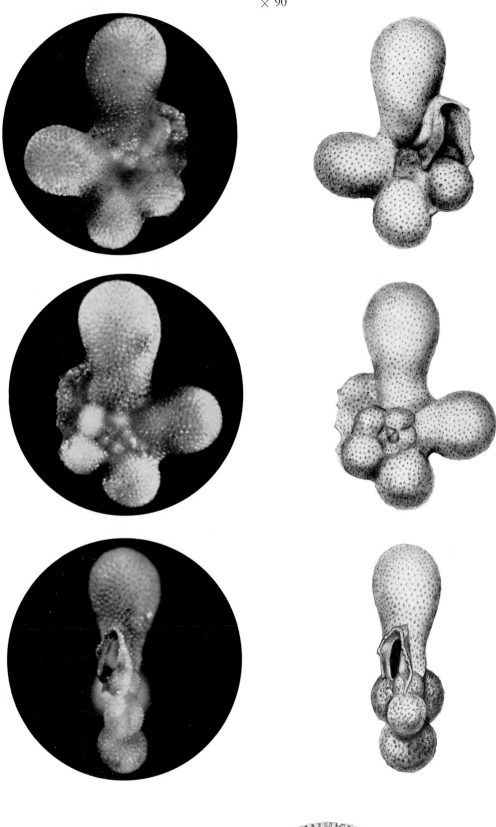

Orbulina bilobata (D'Orbigny)

Reference   *Globigerina bilobata* D'Orbigny, 1846: Foraminifères fossiles du bassin tertiaire de Vienne. — Gide and Comp., Paris, p. 164, pl. 9, figs. 11-14.

Type locality   The surroundings of Nussdorf, north of Vienna, Austria.

Diagnosis   Test bilobate, early stage trochospiral.
Wall distinctly perforate, surface pitted.
Chambers spherical, the penultimate one partially or almost completely embracing the globigerine coil.
Primary aperture interiomarginal, umbilical in the early globigerine stage; in the adult areal with numerous small openings, which are scattered over the wall of the last chamber; small sutural secondary openings may be found along the sutures.

Strat. distr.   Base *Globorotalia peripheroronda* zone to Recent

Remarks   See remarks *Globigerinoides sicanus* De Stefani.
Locality of figured specimen is S-13, Italy.

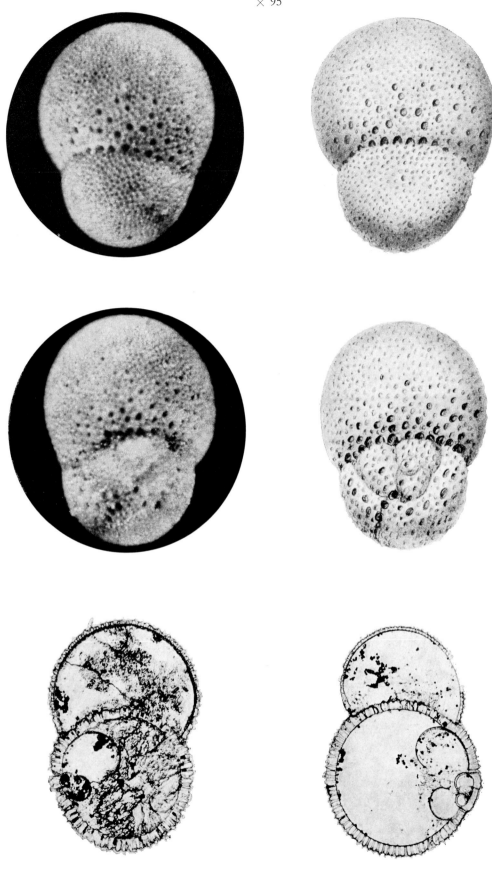

Orbulina suturalis Bronnimann

Reference   *Orbulina suturalis* Bronnimann, 1951: The genus *Orbulina* D'Orbigny in the Oligo-Miocene of Trinidad, B.W.I. — Contributions from the Cushman Foundation for Foraminiferal Research, 2 (4):135, text. fig. IV, figs. 15, 16, 20.

Type locality   Trinidad Leaseholds Ltd. sample no. 160637, from the Naparima area, Trinidad, W.I.

Diagnosis   Test almost globular, early stage trochospiral.
Wall distinctly perforate, surface moderately pitted.
Chambers spherical; the much inflated final chamber not entirely enveloping the early part of the test; transitions from forms in which the last chamber envelops about 75 percent of the earlier part to forms in which the earlier chambers are only just visible have been observed.
Primary aperture interiomarginal, umbilical in the early globigerine stage; in the adult areal with small openings, which are scattered over the wall of the last chamber; small sutural secondary openings are present in the suture separating the last chamber from the penultimate and earlier chambers.

Strat. distr.   Base *Globorotalia peripheroronda* zone to Recent.

Remarks   See remarks *Globigerinoides sicanus* De Stefani.
Locality of figured specimen is TW 623, Trinidad, W.I.

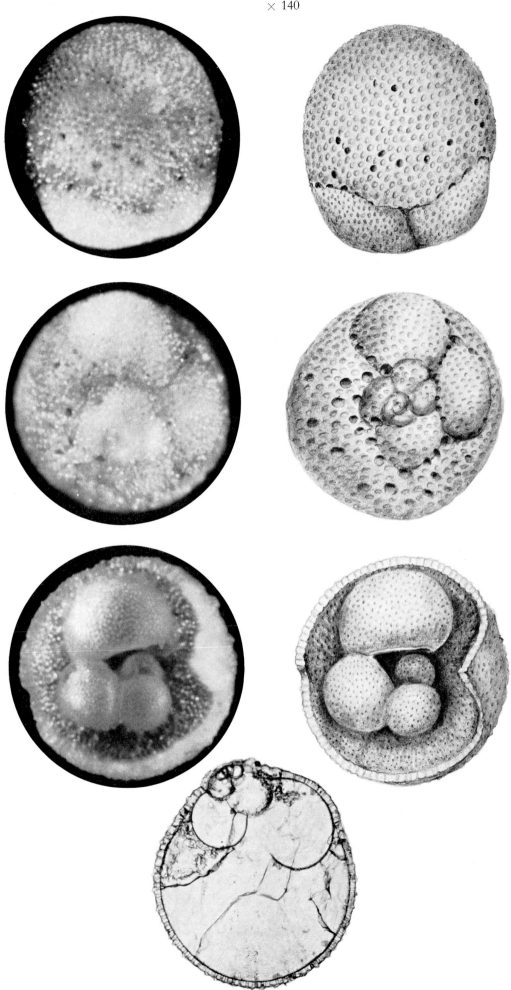

Orbulina universa D'Orbigny

Reference *Orbulina universa* D'Orbigny, 1839: Foraminifères. In: De La Sagra, Histoire physique, politique et naturelle de l'Ile de Cuba. — 8:3, pl. 1, fig. 1.

Type locality Not designated. Localities given: Adriatic Sea near Rimini, Italy; the coast of Algeria; etc.

Diagnosis Test globular, early stage trochospiral, globigerine-like.
Wall distinctly perforate, surface moderately pitted.
Chambers spherical, the globular final chamber entirely enveloping the early part of the test, which is usually reduced.
Primary aperture interiomarginal, umbilical in the early stage; in the adult areal with numerous small openings, which are scattered over the wall of the last chamber.

Strat. distr. Base *Globorotalia peripheroronda* zone to Recent.

Remarks See remarks *Globigerinoides sicanus* De Stefani.
Locality of figured specimen is S 10, Italy.

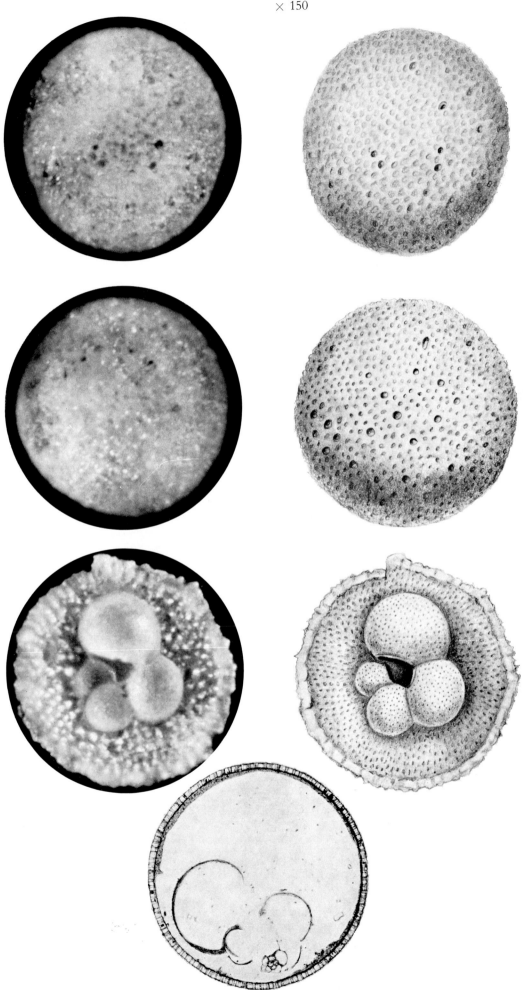

Praeorbulina glomerosa (Blow)

Reference *Globigerinoides glomerosa* Blow, 1956: Origin and evolution of the foraminiferal genus *Orbulina* D'Orbigny. — Micropaleontology, 2 (1):64, textfig. 1, nos. 9-19, textfig. 2, nos. 1-4.

Type locality An auger line near Pozón, District of Acosta, eastern Falcón, Venezuela.

Diagnosis Test subglobular, ovoid to nearly spherical, early portion trochospiral; periphery almost circular.
Wall distinctly perforate, surface slightly pitted.
Chambers spherical, arranged in about four whorls; the last chambers increase rapidly in size; the final chamber much inflated and embracing the earlier part of the test up to more than 40 percent.
Sutures slightly curved to radial, slightly depressed.
Primary aperture in the early portion interiomarginal, umbilical, with secondary sutural openings on the spiral side, as in *Globigerinoides*. The primary aperture of the early portion is covered by the final embracing chamber, which has only numerous (four to twenty) small, sutural, secondary apertures encircling its basal margin.

Strat. distr. Uppermost part *Globigerinatella insueta* zone into upper part *Globorotalia peripheroronda* zone.

Remarks Blow (1956) published his study "Origin and evolution of the foraminiferal genus *Orbulina* D'Orbigny" (Micropal., 2:57-70).
He distinguished a.o. the following three subspecies:
Globigerinoides glomerosa curva
Globigerinoides glomerosa glomerosa
Globigerinoides glomerosa circularis
As the evolution of this series took place in a very short time interval and their identification is only possible if very well preserved material is available, we only recognize, for practical purpose, the species *Praeorbulina glomerosa* (Blow) as a useful index fossil. See remarks *Globigerinoides sicanus* De Stefani.
Locality of figured specimen is RD-69, Falcón, Venezuela.

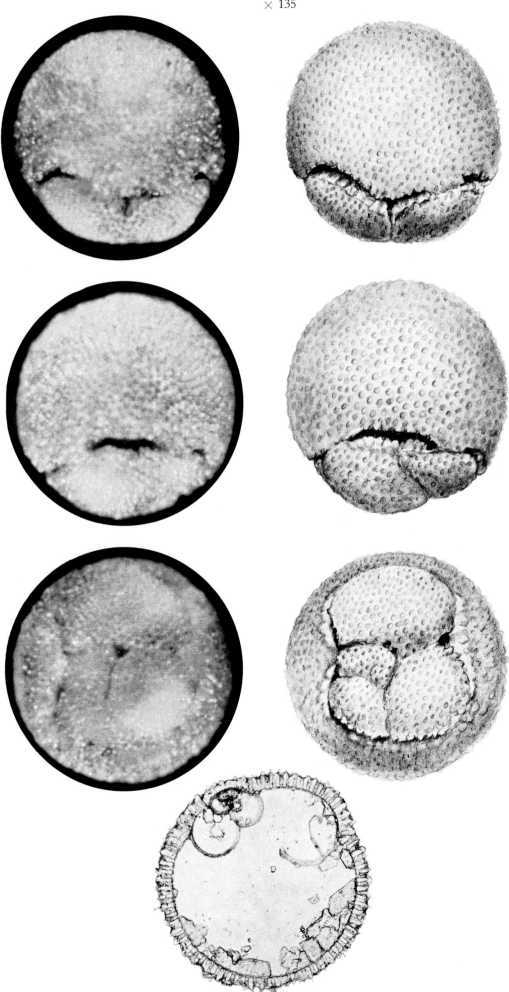

Praeorbulina transitoria (BLOW)

Reference   *Globigerinoides transitoria* BLOW, 1956: Origin and evolution of the foraminiferal genus *Orbulina* D'ORBIGNY. — Micropaleontology, 2 (1):65, textfig. 2, nos. 12-13.

Type locality   An auger line near Pozón, District of Acosta, eastern Falcón, Venezuela.

Diagnosis   Early portion of test trochospiral; equatorial periphery bilobate; axial periphery rounded. Wall distinctly perforated, surface pitted.
Chambers spherical, arranged in three and a half to four whorls; the penultimate and ultimate chambers nearly equal in size, embracing the early chambers.
Sutures on spiral side slightly curved to radial, slightly depressed.
Primary aperture in the early portion interiomarginal, umbilical, with secondary sutural small openings on the spiral side, as in *Globigerinoides*; the final chamber has three or more sutural secondary apertures, low arches or narrow slits, confined to the suture of the last chamber.

Strat. distr.   Upper part of *Globigerinatella insueta* zone into lower part *Globorotalia peripheroronda* zone.

Remarks   See remarks *Globigerinoides sicanus* DE STEFANI.
Locality of figured specimen is RD-22, Mirimire-Tucacas Road, Falcón, Venezuela.

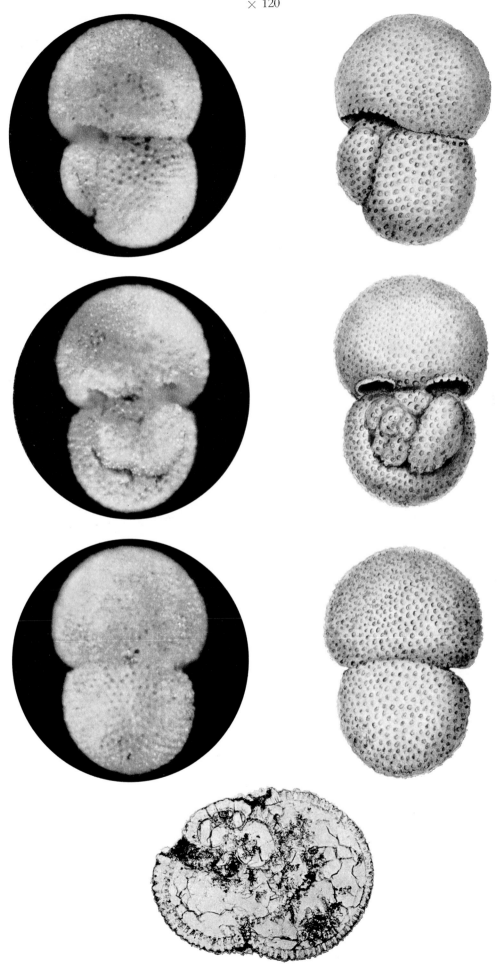

Pulleniatina obliquiloculata (PARKER and JONES)

Reference *Pullenia sphaeroides* (D'ORBIGNY) var. *obliquiloculata* PARKER and JONES, 1865: On some foraminifera from the North Atlantic and Arctic Oceans, including Davis Straits and Baffin's Bay. — Roy. Soc. London, Philos. Trans., 155:368, pl.19, fags. 4a-b.

Type locality Not designated. Localities given: Abrolhos Bank, lat. 22°54'S., long. 40°37'W., Atlantic Ocean; lat. 2°20'N., long. 48°44'W., tropical Atlantic Ocean; lat. 5°37'S., long. 61°33' E., Indian Ocean.

Diagnosis Test globose, early stage trochospiral, final stage streptospiral, almost involute; equatorial periphery slightly lobulate; axial periphery broadly rounded.
Wall finely perforate, wall of early whorls on spiral side coarser perforate and thicker; surface smooth, except for a granular area immediately facing the aperture.
Chambers subspherical, arranged in about three whorls; the last whorl consisting of four and a half to five chambers, all of them increasing the degree of embrace, due to a deviation of the mode of coiling, forming a coil which is partially or almost completely involute.
Sutures on spiral side slightly curved to almost radial, depressed.
Umbilicus covered.
Aperture a fairly low to medium high arch along the base of the final chamber, reaching farther than the periphery of the preceding whorl, bordered above by a hyeline "rim", which may possess scattered granules.

Strat. distr. Uppermost part *Globoquadrina altispira* zone to Recent.

Remarks The easiest way to distinguish *Pulleniatina primalis, P. praecursor* and *P. obliquiloculata* from one another is to compare the position of the apertures. The aperture of *P. primalis* is confined to the umbilical part of the final basal suture, not reaching the periphery of the preceding whorl. The aperture of *P. praecursor* just reaches the periphery, while the aperture of *P. obliquiloculata* reaches at least farther than the periphery of the preceding whorl, in recent specimens even becoming almost "equatorial" in position.
Locality of figured specimen is Challenger Station 224, north of Admiralty Islands, Pacific Ocean.

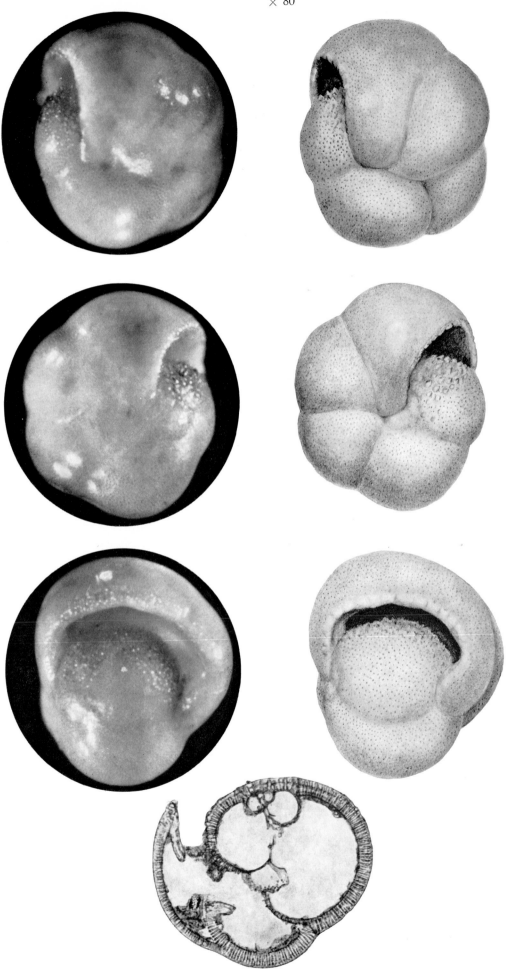

Pulleniatina praecursor BANNER and BLOW

Reference   *Pulleniatina obliquiloculata* (PARKER and JONES) *praecursor* BANNER and BLOW: The Origin, Evolution and Taxonomy of the Foraminiferal Genus *Pulleniatina,* Cushman 1927. – Micropaleontology, 13 (2): 139, pl. 3, Figs. 3a-c.

Type locality   Stainforth sample 19305, between Rio Chevele and Estero Ciénago, 345 km. north and 13 km. east of the Plaza del Centenario monument in Guayaquil, Ecuador.

Diagnosis   Test low to medium high trochospiral, except for the last chambers which are more embracing; equatorial periphery lobulate; axial periphery broadly rounded.
Wall finely perforate, wall of early whorls on spiral side coarser perforate and thicker; surface smooth, except for a granular area immediately facing the aperture.
Chambers subspherical, arranged in about three whorls; the second whorl consisting of four chambers, the last whorl consisting of five chambers, all chambers increase regularly in size except the last ones, which abruptly increase the degree of embrace due to a deviation of the mode of coiling.
Sutures on spiral side slightly curved, depressed; on umbilical side almost radial, depressed.
Umbilicus covered by the last chamber.
Aperture interiomarginal, extraumbilical-umbilical, a long, very low arch along the base of the final chamber, just reaching the periphery of the preceding whorl, bordered above by a hyaline "rim"

Strat. distr.   Lower part *Globoquadrina altispira* zone into lower part *Globorotalia tosaensis* zone. Questionable occurrences in uppermost part *Globorotalia margaritae* zone, lowermost part *G. altispira* zone and upper part *G. tosaensis* zone.

Remarks   See *Pulleniatina obliquiloculata* (PARKER and JONES).
Locality of figured specimen is well Bodjonegoro - 1, 221m., Java., Indonesia.

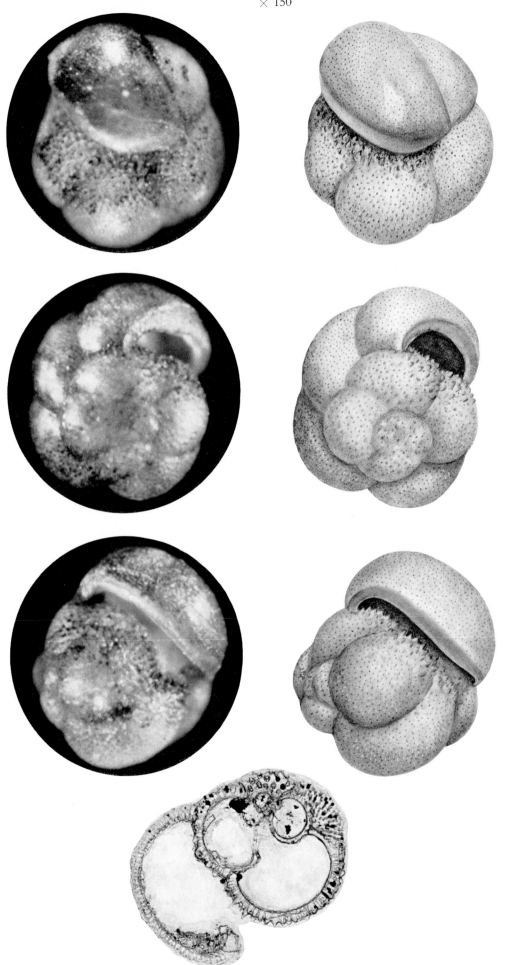

Pulleniatina primalis BANNER and BLOW

Reference   *Pulleniatina primalis* BANNER and BLOW: The Origin, Evolution and Taxonomy of the Foraminiferal Genus *Pulleniatina*, CUSHMAN 1927. – Micropaleontology, 13 (2): 142, pl. 3, figs. 2a-c.

Type locality   Sample WHB. 181. B, Buff Bay, Jamaica, W.I.

Diagnosis   Test low to medium high trochospiral; equatorial periphery lobulate; axial periphery broadly rounded.

Wall finely perforate; surface smooth, except for a granular area immediately facing the aperture.

Chambers subspherical, arranged in about three whorls; the five chambers of the last whorl increase regularly in size except the final one, which abruptly increases the degree of embrace due to a deviation of the mode of coiling.

Sutures on spiral side slightly curved, depressed; on umbilical side almost radial, depressed.

Umbilicus just covered by the last chamber.

Aperture interiomarginal, extraumbilical-umbilical, a broad, low arch along the base of the final chamber, not reaching the periphery of the preceding whorl, bordered above by a hyaline "rim".

Strat. distr.   Uppermost part *Globorotalia dutertrei* zone to top *Globoquadrina altispira* zone. Questionable occurrences in middle part *Globorotalia dutertrei* zone and lowermost part *G. tosaensis* zone.

Remarks   See *Pulleniatina obliquiloculata* (PARKER and JONES).
Locality of figured specimen is Araya, Estado Sucre, Venezuela.

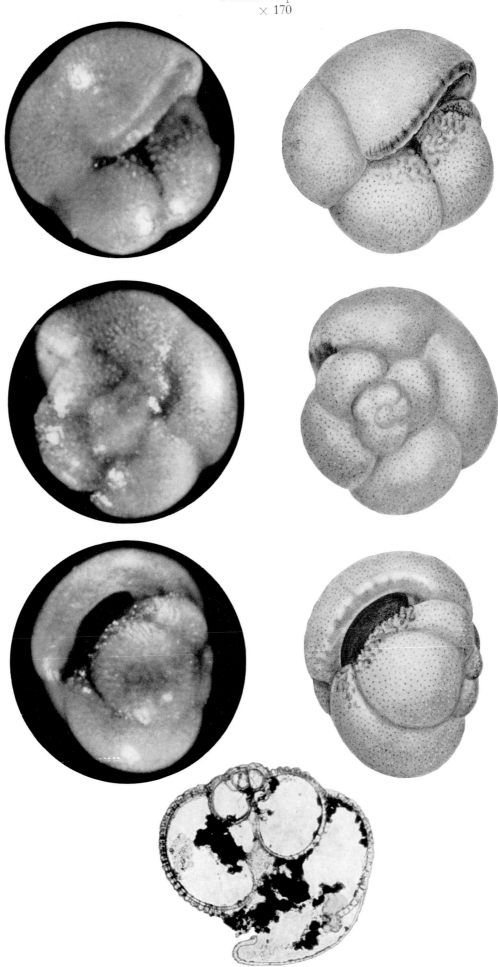

Sphaeroidinella dehiscens (PARKER and JONES)

Reference *Sphaeroidina bulloides* D'ORBIGNY var. *dehiscens* PARKER and JONES, 1865: On some Foraminifera from the North Atlantic and Arctic Oceans, including Davis Straits and Baffin's Bay. — Roy. Soc. London, Philos. Trans., 155:369, pl.19, fig. 5.

Type locality Not designated. Localities given: 1080 fathoms, lat. 2° 20' N., long. 28° 44' W., tropical Atlantic Ocean; 2200 fathoms, lat. 5° 37' S., long. 61° 33' E., Indian Ocean.

Diagnosis Test trochospiral, compact; equatorial periphery very slightly lobulate; axial periphery rounded.
Primary wall coarsely perforate, covered by secondary layers of shell material greatly reducing the external openings of the pores of the primary wall or sealing them; surface smooth and glassy in appearance.
Chambers subglobular, becoming increasingly embracing in the adult, arranged in about three whorls; the three chambers of the last whorl increase rapidly in size.
Sutures indistinct, radial, slightly depressed.
Primary aperture interiomarginal, umbilical; one or two sutural secondary apertures on opposite side of the final chamber are present. Primary as well as secondary apertures may be partially obscured by the smooth or crenulate overhanging chamber flanges.

Strat. distr. Uppermost part *Globorotalia margaritae* zone to Recent. Questionable occurrence in middle part *G. margaritae* zone.

Remarks See remarks *Sphaeroidinella subdehiscens* BLOW.
Locality of figured specimen is 524 "Downwind", BG 137, lat. 9° 53' S., long. 110° 41' W., Pacific Ocean.

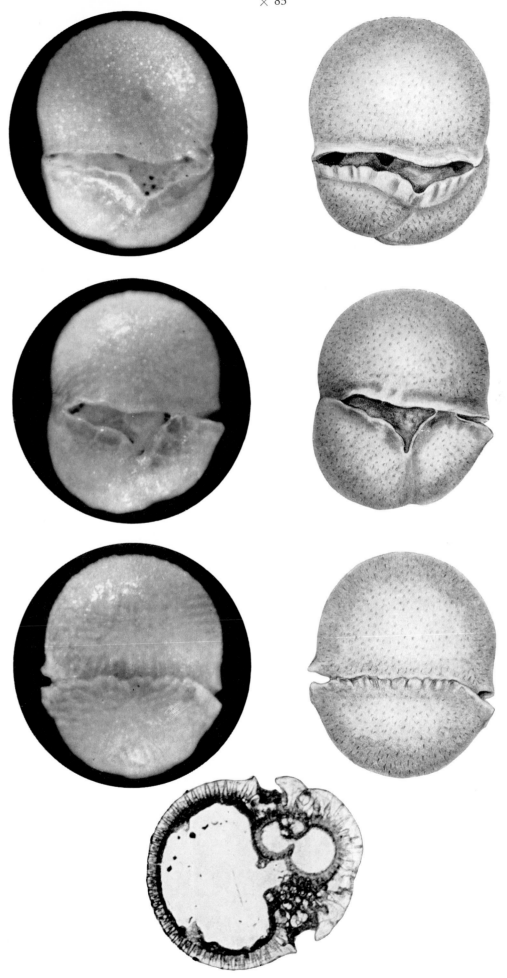

Sphaeroidinella subdehiscens BLOW

Reference *Sphaeroidinella dehiscens* (PARKER and JONES) *subdehiscens* BLOW, 1959: Age, correlation, and biostratigraphy of the Upper Tocuyo (San Lorenzo) and Pozón Formations, eastern Falcón, Venezuela. – Bulletins of American Paleontology, 39 (178): 195, pl. 12, figs. 71a-c.

Type locality Sample RM 19514, auger line near Pozón, eastern Falcón, Venezuela.

Diagnosis Test low trochospiral, compact; equatorial periphery slightly lobulate; axial periphery rounded.
Primary wall coarsely perforate, covered by secondary layers of shell material greatly reducing the external openings of the pores of the primary wall or sealing them; surface smooth and glassy in appearance.
Chambers subglobular to somewhat radially elongate, rather embracing, arranged in about three whorls; the three chambers of the last whorl increase fairly slowly in size.
Sutures almost straight on spiral and umbilical side, slightly depressed.
Aperture interiomarginal, umbilical, an elongate slit or low arch, with thickened margins, which may be smooth or crenulate.

Strat. distr. Base *Globigerinoides subquadratus* zone to top *Globoquadrina altispira* zone.

Remarks This species is distinguished from *Sphaeroidinella dehiscens* (PARKER and JONES) by its less embracing chambers, more distinctive sutures, absence of supplementary apertures, a slightly more lobulate equatorial periphery and a generally smaller test.
BLOW (1960) created the new genus *Sphaeroidinellopsis* for species as described above without secondary apertures.
Locality of figured specimen is CF 1611, Trinidad, W.I.

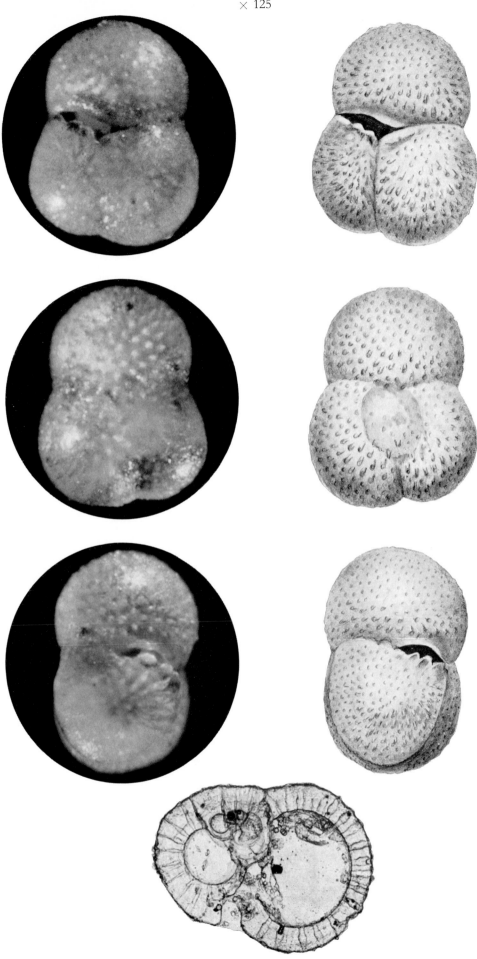

Figure 19 – Foraminiferal packstone with unidentified *Globigerina spp.* and
Globorotalia spp.   × 23
Sample Bz 43, Marches-Umbria area, Italy.
Globigerina ampliapertura / *G. angulisuturalis* zone.

Figure 20 – Foraminiferal pack/wackestone with *Globigerinoides trilobus s.l.* and
Globorotalia kugleri BOLLI.   × 20
Sample O 779, Onin, W. Irian.
Lowermost *Globigerinoides trilobus* zone.

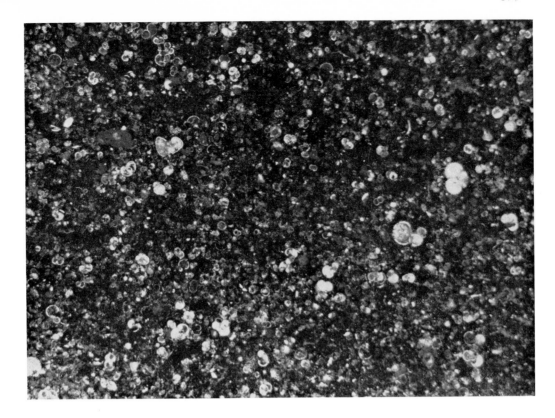

Figure 19

Figure 20

Figure 21 — Foraminiferal wackestone with *Globigerina binaiensis* Koch, *Globorotalia kugleri* Bolli and *Globigerinoides trilobus s.l.*   × 25
Sample Vr 193, Onin, W. Irian.
Lowermost *Globigerinoides trilobus* zone.

393

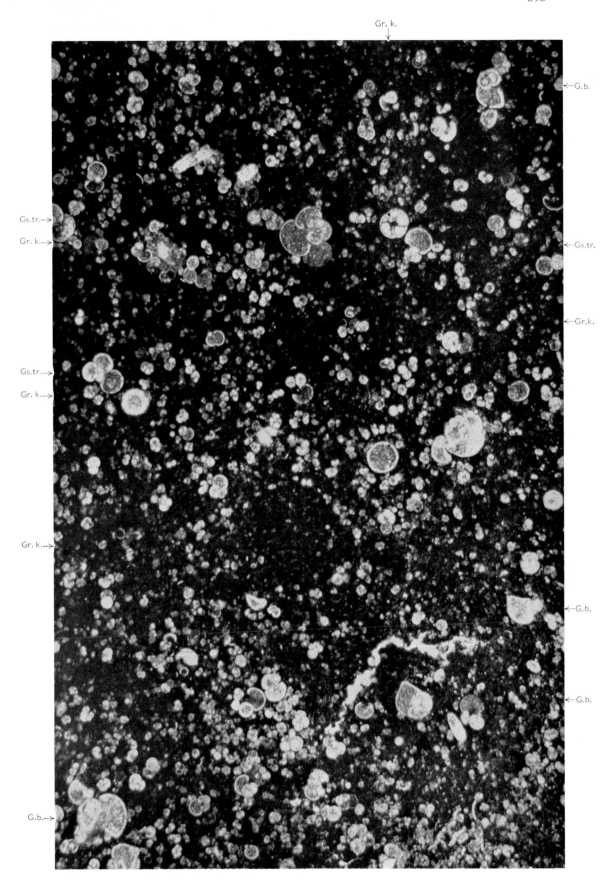

Figure 21

Figure 22 – Foraminiferal pack/wackestone with *Globigerinoides trilobus s.l.*, *Globigerinoides sicanus* DE STEFANI and *Globoquadrina altispira s.l.* ×18
Sample Bz 30, Marches-Umbria area, Italy.
Upper part *Globigerinatella insueta* zone.

Figure 23 — Foraminiferal pack/wackestone with *Globorotalia praemenardii* CUSHMAN and STAINFORTH, *Globorotalia lobata/fohsi* and *Orbulina s.l.* ×24
Sample O 763, Onin, W. Irian.
Globorotalia lobata/fohsi zone.

Figure 22

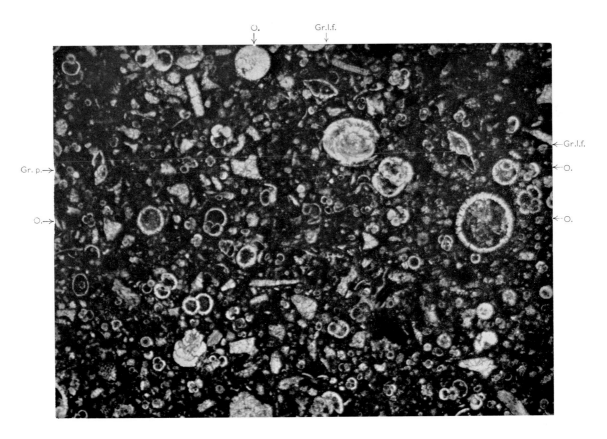

Figure 23

Fig. 24 – Foraminiferal pack/wackestone with *Sphaeroidinella dehiscens* (PARKER and JONES), *Globorotalia crassaformis/tosaensis* group, *Globorotalia tumida* (BRADY) and *Pulleniatina* sp.   ×15
Sample Mu 66, Vaingana, Eua, Tonga Islands.
Probably *Globorotalia tosaensis* zone.

Figure 24

GLOSSARY

Aperture, primary: The main aperture opening to the exterior from the final chamber of the test. May vary in position as follows:
equatorial: symmetrical, just above the peripheral margin of the previous whorl in planispiral or nearly planispiral tests.
extraumbilical–umbilical: extending, in part at least, outside the umbilicus along the forward margin of the final chamber.
umbilical: opening directly into the umbilicus of the test.

Apertures, secondary: Smaller openings in addition to the primary aperture, or remnants of primary apertures.
accessory: secondary apertures which do not open directly into a primary chamber but into or under an accessory structure (i.e. bulla).

Angular conical: Inflated with angular margins and conical in shape.

Angular rhomboid: Rhombic in section and sharply angled.

Angular truncate: Inflated with truncate margins, commonly keeled.

Beaded: Provided with ornamentations like a string of beads.

Bulla: A perforate, inflated, accessory structure which covers the umbilicus (and sometimes the supplementary apertures); one or more accessory apertures are present at the margin.

Carina: An imperforate, ridge-like thickening of the chamber wall, present on the periphery of the test, lying in the plane of coiling.

Clavate: Elongated, may be inflated terminally, having a club-shaped appearance.

Enrolled biserial: A planispiral development in which biserially alternating chambers are enrolled.

Flange, apertural: A rather broad lip, found on both sides of an elevated, peripheral aperture.

Flange, chamber: A broad fold developed along the basal margin of a chamber that tends to obscure the suture.

Flap: An imperforate, fairly broad structure projecting from, and apparently additional to, the chamber wall, above and along a primary aperture.

Flush: Even or level, forming a continuous surface.

Hemispherical: Inflated on one side, flattened on the opposite side.

Hispid: Characterized by very fine, short, hair-like spines.

Infralaminal: Along the margin of an accessory structure (i.e. bulla).

Interiomarginal: At the base of the final chamber.

Intralaminal: Piercing an accessory structure (i.e. tegillum).

Keel: An imperforate, ridge-like thickening of the chamber wall, present on the periphery of the test, lying in the plane of coiling.

Limbate: Bordered by or provided with a flat or raised strip.

Lip: A fairly narrow structure which appears to be a reflexed continuation of the actual chamber wall, projecting above and along an interiomarginal aperture; typically symmetrically developed about the aperture.

Nodose: Surface ornamented with small knobs.

Pitted: Small, generally rounded depressions in the surface of the wall.

Pseudocarina (pseudokeel): A perforate, ridge-like thickening of the peripheral part of the chamber wall, lying approximately in the plane of coiling.

Reticulate: With a honeycomb-like surface.

Rugose: Characterized by rough ornamentation, which may form ridges.

Spinose: Provided with very fine spines, generally elongate.

Spiral side (dorsal): The more evolute side of a trochospirally coiled test.

Stellate: Star-shaped.

Streptospiral: A modification of the trochospiral coiling in which the plane of coiling continually changes.

Tegillum: A complex, often irregular structure of imperforate plates which covers the umbilicus; the later-formed plates may extend the full breadth of the umbilicus and fuse on to the walls of the chambers opposite. The tegillum may have smaller openings along the margin, or be pierced (infralaminal and intralaminal apertures).

Truncate: Having the ends square or even, as though cut off.

Tubulospine: A long, hollow chamber extension.

Umbilical side (ventral): The more involute side of a trochospirally coiled test, typically possessing an umbilicus.

Umbilicate: Having one or more umbilici.

REFERENCES

AKERS, W. H., 1955. Some planktonic foraminifera of the American Gulf Coast and suggested correlations with the Caribbean Tertiary. *J. Paleontol.*, 29(4): 647–664.

AKERS, W. H. and DROOGER, C. W., 1957. Miogypsinids, planktonic foraminifera and Gulf Coast Oligocene–Miocene correlations. *Bull. Am. Assoc. Petrol. Geologists*, 41(4): 656–678.

APPLIN, E. R., 1964. Some Middle Eocene, Lower Eocene, and Paleocene foraminiferal faunas from West Florida. *Contrib. Cushman Found. Foram. Res.*, 15(2): 45–72.

BANDY, O. L., 1949. Eocene and Oligocene Foraminifera from Little Stave Creek, Clarke County, Alabama. *Bull. Am. Paleontol.*, 32(131): 5–240.

—, 1963. Cenozoic planktonic foraminiferal zonation and basinal development in the Philippines. *Bull. Am. Assoc. Petrol. Geologists*, 47(9): 1733–1745.

—, 1967. Cretaceous planktonic foraminiferal zonation. *Micropaleontology*, 13(1): 1–31.

BANDY, O. L. and INGLE, J. C., 1970. Neogene planktonic events and radiometric scale, California. *Geol. Soc. Am., Spec. Papers*, 124: 131–172.

BANDY, O. L. and WADE, M., 1967. Miocene–Pliocene–Pleistocene boundaries in deep-water environments. *Progr. Oceanography*, 4: 51–66.

BANNER, F. T. and BLOW, W. H., 1959. The classification and stratigraphical distribution of the Globigerinaceae. *Palaeontology*, 2(1): 1–27.

—, 1960. Some primary types of species belonging to the superfamily Globigerinaceae. *Contrib. Cushman Found. Foram. Res.*, 11(1): 1–41.

—, 1965a. Progress in the planktonic foraminiferal biostratigraphy of the Neogene. *Nature*, 208(5016): 1164–1166.

—, 1965b. Two new taxa of the Globorotaliinae assisting determination of the Late Miocene/Middle Miocene boundary. *Nature*, 207(5004): 1351–1354.

—, 1965c. *Globigerinoides quadrilobatus* (D'ORBIGNY) and related forms: their taxonomy, nomenclature and stratigraphy. *Contrib. Cushman Found. Foram. Res.*, 16(3): 105–115.

—, 1967. The origin, evolution and taxonomy of the foraminiferal genus *Pulleniatina* CUSHMAN, 1927. *Micropaleontology*, 13(2): 133–162.

BANNER, F. T. and EAMES, F. E., 1966. Recent progress in world-wide Tertiary stratigraphical correlation. *Earth-Sci. Rev.*, 2(2): 157–179.

BARNARD, T., 1954. *Hantkenina alabamensis* CUSHMAN and some related forms. *Geol. Mag.*, 91(5): 384–390.

BATJES, D. A. J., 1958. Foraminifera of the Oligocene of Belgium. *Mém. Inst. Sci. Natl. Belg.*, 143: 1–88.

BAYLISS, D. D., 1969. The distribution of *Hyalinea balthica* and *Globorotalia truncatulinoides* in the type Calabrian. *Lethaia*, 2(2): 133–143.

BE, A. W. H., 1965. The influence of depth on shell growth in *Globigerinoides sacculifer* (BRADY). *Micropaleontology*, 11(1): 81–97.

BECKMANN, J. P., 1953. Die Foraminiferen der Oceanic Formation (Eocaen–Oligocaen) von Barbados, Kleine Antillen. *Eclogae Geol. Helv.*, 46(2): 301–412.

—, 1958. Correlation of pelagic and reefal faunas from the Eocene and Paleocene of Cuba. *Eclogae Geol. Helv.*, 51(2): 416–421.

BECKMANN, J. P., EL-HEINY, I., KERDANY, M. T., SAID, R. and VIOTTI, C., 1969. Standard planktonic zones in Egypt. *Proc. Intern. Conf. Planktonic Microfossils, 1st*, 1: 92–103.

BERGGREN, W. A., 1960a. Some planktonic Foraminifera from the Lower Eocene (Ypresian) of Denmark and northwestern Germany. *Stockholm Contrib. Geol.*, 5(3): 41–108.

—, 1960b. Biostratigraphy, planktonic Foraminifera and the Cretaceous–Tertiary boundary in Denmark and southern Sweden. *Intern. Geol. Congr., 21st, Copenhagen, 1960. Sect. 5*, pp.181–192.

—, 1962. Some planktonic Foraminifera from the Maestrichtian and type Danian stages of southern Scandinavia. *Stockholm Contrib. Geol.*, 9(1): 1–106.

—, 1969. Cenozoic chronostratigraphy, planktonic foraminiferal zonation and the radiometric time scale. *Nature*, 224(5224): 1072–1075.

BERMUDEZ, P. J., 1949. Tertiary smaller Foraminifera of the Dominican Republic. *Cushman Lab. Foram. Res. Spec. Publ.*, 25: 1–322.

—, 1952. Estudio sistematico de los foraminiferos rotaliformes. *Venezuela, Min. Minas Hidrocarburos, Bol. Geol.*, 2(4): 1–230.

—, 1958. Los Foraminiferos pelagicos de la region Caribe-Antillana (Paleoceno–Reciente). *Acta Sci. Venezolana*, 9(6,7): 122–125.

—, 1960. Contribución al estudio de las Globigerinidea de la región Caribe–Antillana (Paleoceno–Reciente). *Bol. Geol., Publ. Esp.*, 3: 1119–1393.

BERMUDEZ, P. J. and BOLLI, H. M., 1969. Consideraciones sobre los sedimentos del Miocene medio al Reciente de las costas central y oriental de Venezuela. *Bol. Geol.*, 10(20): 137–224.

BERTOLINO, V. et al., 1968. Proposal for a biostratigraphy of the Neogene in Italy based on planktonic Foraminifera. *Giorn. Geol., Ann. Museo Geol. Bologna*, 35(2): 23–30.

BLOW, W. H., 1956. Origin and evolution of the foraminiferal genus *Orbulina* D'ORBIGNY. *Micropaleontology*, 2(1): 57–70.

—, 1957. Trans-Atlantic correlation of Miocene sediments. *Micropaleontology*, 3(1): 77–79.

—, 1959. Age, correlation and biostratigraphy of the Upper Tocuyo (San Lorenzo) and Pozón Formations, eastern Falcón, Venezuela. *Bull. Am. Paleontol.*, 39(178): 1–251.

—, 1969. Late Middle Eocene to Recent planktonic foraminiferal biostratigraphy. *Proc. Intern. Conf. Planktonic Microfossils, 1st*, 1: 199–421.

BLOW, W. H. and BANNER, F. T., 1962. The Mid-Tertiary (Upper Eocene to Aquitanian) Globigerinaceae. In: F. E. EAMES et al., *Fundamentals of Mid-Tertiary Stratigraphical Correlation*. Cambridge University Press, London, pp.61–151.

—, 1966. The morphology, taxonomy and biostratigraphy of *Globorotalia barisanensis* LEROY, *Globorotalia fohsi* CUSHMAN and ELLISOR, and related taxa. *Micropaleontology*, 12(3): 286–302.

BLOW, W. H. and SAITO, T., 1968. The morphology and taxonomy of *Globigerina mexicana* CUSHMAN, 1925. *Micropaleontology*, 14(3): 357–360.

BOLLI, H. M., 1944. Zur Stratigraphie der Oberen Kreide in den höheren helvetischen Decken. *Eclogae Geol. Helv.*, 37(2): 217–328.

—, 1951. The genus *Globotruncana* in Trinidad, B.W.I. *J. Paleontol.*, 25(2): 187–199.

—, 1954. Note on *Globigerina concinna* REUSS, 1850. *Contrib. Cushman Found. Foram. Res.*, 5(1): 1–3.

—, 1957a. The genera *Praeglobotruncana*, *Rotalipora*, *Globotruncana* and *Abathomphalus* in the Upper Cretaceous of Trinidad, B.W.I. *U.S. Natl. Museum Bull.*, 215: 51–60.

—, 1957b. The genera *Globigerina* and *Globorotalia* in the Paleocene–lower Eocene Lizards Springs Formation of Trinidad, B.W.I. *U.S. Natl. Museum Bull.*, 215: 61–81.

—, 1957c. Planktonic Foraminifera from the Eocene Navet and San Fernando Formations of Trinidad, B.W.I. *U.S. Natl. Museum Bull.*, 215: 155–172.

—, 1957d. Planktonic Foraminifera from the Oligocene–Miocene Cipero and Lengua Formations of Trinidad, B.W.I. *U.S. Natl. Museum Bull.*, 215: 97–123.

—, 1957e. The foraminiferal genera *Schackoina* THALMANN, emended, and *Leupoldina* n. gen. in the Cretaceous of Trinidad, B.W.I. *Eclogae Geol. Helv.*, 50(2): 271–278.

—, 1959a. Planktonic Foraminifera from the Cretaceous of Trinidad, B.W.I. *Bull. Am. Paleontol.*, 39(179): 257–277.

—, 1959b. Planktonic Foraminifera as index fossils in Trinidad, West-Indies and their value for worldwide stratigraphic correlation. *Eclogae Geol. Helv.*, 52(2): 627–637.

—, 1964. Observations on the stratigraphic distribution of some warm-water planktonic foraminifera in the young Miocene to Recent. *Eclogae Geol. Helv.*, 57(2): 541–552.

—, 1966a. The planktonic foraminifera in well Bodjonegoro-1 of Java. *Eclogae Geol. Helv.*, 59(1): 449–465.

—, 1966b. Zonation of Cretaceous to Pliocene marine sediments based on planktonic Foraminifera. *Bol. Inform. Asoc. Venezolana Geol. Mineria Petrol.*, 9(1): 3–32.

BOLLI, H. M. and BERMUDEZ, P. J., 1965. Zonation based on planktonic Foraminifera of Middle Miocene to Pliocene warm-water sediments. *Bol. Inform. Asoc. Venezolana Geol. Mineria Petrol.*, 8(5): 119–149.

BOLLI, H. M. and CITA, M. B., 1960a. Globigerine e Globorotalie del Paleocene di Paderno d'Adda (Italia). *Riv. Ital. Paleontol. Stratigraf.*, 66(3): 361–402.

—, 1960b. Upper Cretaceous and Lower Tertiary planktonic Foraminifera from the Paderno d'Adda section, northern Italy. *Intern. Geol. Congr., 21st, Copenhagen, 1960, Sect. 5*, pp.150–161.

BOLLI H. M., CITA, M. B. and SCHAUB, H., 1962. Il limite Cretaceo–Terziario nella catena del Monte Baldo. *Mem. Soc. Geol. Ital.*, 3: 149–168.

BOLLI, H. M., LOEBLICH, A. R. and TAPPAN, H., 1957. Planktonic foraminiferal families Hantkeninidae, Orbulinidae, Globorotaliidae, and Globotruncanidae. *U.S. Natl. Museum Bull.*, 215: 3–60.

BORSETTI, A. M., 1959. Tre nuovi Foraminiferi planctonici dell' Oligocene–Piacentino. *Giorn. Geol., Ann. Museo Geol. Bologna*, 27(2a): 205–212.

BRADSHAW, J. S., 1959. Ecology of living planktonic Foraminifera in the North and Equatorial Pacific Ocean. *Contrib. Cushman Found. Foram. Res.*, 10(2): 25–64.

BRADY, H. B., 1884. Report on the Foraminifera dredged by H.M.S. Challenger during the years 1873–1876. *Rept. Challenger Expedition, Zool.*, 9(22): 1–814.

BRÖNNIMANN, P., 1950a. The genus *Hantkenina* CUSHMAN in Trinidad and Barbados, B.W.I. *J. Paleontol.*, 24(4): 397–420.

—, 1950b. Weitere Beobachtungen an Hantkeninen. *Eclogae Geol. Helv.*, 43(2): 245–251.

—, 1951a. *Globigerinita naparimaensis* n. gen., n. sp., from the Miocene of Trinidad, B.W.I. *Contrib. Cushman Found. Foram. Res.*, 2(1): 16–18.

—, 1951b. The genus *Orbulina* D'ORBIGNY in the Oligo–Miocene of Trinidad, B.W.I. *Contrib. Cushman Found. Foram. Res.*, 2(4): 132–138.

—, 1952a. Globigerinidae from the Upper Cretaceous (Cenomanian–Maestrichtian) of Trinidad, B.W.I. *Bull. Am. Paleontol.*, 34(140): 1–70.

—, 1952b. Trinidad Paleocene and Lower Eocene Globigerinidae. *Bull. Am. Paleontol.*, 34(143): 1–34.

—, 1952c. *Plummerita* new name for *Plummerella* BRÖNNIMANN, 1952. *Contrib. Cushman Found. Foram. Res.*, 3(4): 146.

—, 1953. Note on planktonic Foraminifera from Danian localities of Jutland, Denmark. *Eclogae Geol. Helv.*, 45(2): 339–341.

BRÖNNIMANN, P. and BROWN JR., N. K., 1955. Taxonomy of Globotruncanidae. *Eclogae Geol. Helv.*, 48(2): 503–561.

—, 1958. *Hedbergella*, a new name for a Cretaceous planktonic foraminiferal genus. *J. Wash. Acad. Sci.*, 48(1): 15–17.

BRÖNNIMANN, P. and TODD, R., 1954. In: R. TODD, P. E. CLOUD, D. LOW and R. G. SCHMIDT, Probable occurrence of Oligocene on Saipan. *Am. J. Sci.*, 252: 680–682.

BROTZEN, F., 1934. Foraminiferen aus dem Senon Palästinas. *Z. Deut. Palästina Ver.*, 57: 20–72.

—, 1936. Foraminiferen aus dem schwedischen untersten Senon von Eriksdal in Schonen. *Sveriges Geol. Undersökn., Arsbok, Ser. C, Avhandl. Uppsat.*, 396, 30(3): 1–206.

CARSEY, D. O., 1926. Foraminifera of the Cretaceous of central Texas. *Univ. Texas Bull.*, 2612: 1–56.

CASTANARES, A. A., 1954. El genero *Globotruncana* CUSHMAN 1927 y su importancia en estratigrafia. *Bol. Asoc. Mex. Geol. Petrol.*, 6(11–12): 353–474.

CITA, M. B., 1948. Ricerche stratigrafiche e micropaleontologiche sul Cretacico e sull 'Eocene di Tignale (Lago di Garda). *Riv. Ital. Paleontol. Stratigraf.*, 54(2): 49–74.

—, 1955. The Cretaceous–Eocene boundary in Italy. *Proc. World Petrol. Congr., 4th, Sect. I/D, Rome.*

—, 1958. Introduzione allo studio del Cretaceo italiano. *Riv. Ital. Paleontol. Stratigraf.*, 64(1): 3–25.

CITA, M. B. and BOLLI, H. M., 1961. Nuovi dati sull'eta' Paleocenica dello Spilecciano di Spilecco. *Riv. Ital. Paleontol. Stratigraf.*, 57(4): 369–392.

CITA, M. B. and SILVA, I. P., 1960. Pelagic foraminifera from the type Langhian. *Intern. Geol. Congr., 21st, Copenhagen, Sect. 22*, pp.39–50.

COLE, W. S., 1927. A foraminiferal fauna from the Guayabal Formation in Mexico. *Bull. Am. Paleontol.*, 14(51): 1–46.

—, 1928. A foraminiferal fauna from the Chapapote Formation in Mexico. *Bull. Am. Paleontol.*, 14(53): 1–32.

COLOM, G., 1954. Estudio de las biozonas con foraminiferos del Terciario de Alicante. *Bol. Inst. Geol. Minero España*, 66: 1–279.

CONATO, V. and FOLLADOR, U., 1967. *Globorotalia crotonensis* e *Globorotalia crassacrotonensis* nuove specie del Pliocene Italiano. *Boll. Soc. Geol. Ital.*, 86: 555–563.

CORDEY, W. G., 1967. The development of *Globigerinoides ruber* (D'ORBIGNY), 1839 from the Miocene to Recent. *Palaeontology*, 10(4): 647–659.

CRESCENTI, U., 1966. Sulla biostratigrafia del Miocene affiorante al confine Marchigiano–Abruzzese. *Geol. Romana*, 5: 1–54.

CUSHMAN, J. A., 1925. New Foraminifera from the Upper Eocene of Mexico. *Contrib. Cushman Lab. Foram. Res.*, 1(1): 4–9.

—, 1926. Some Foraminifera from the Mendez Shale of eastern Mexico. *Contrib. Cushman Lab. Foram. Res.*, 2(1): 16–26.

—, 1927a. An outline of a reclassification of the Foraminifera. *Contrib. Cushman Lab. Foram. Res.*, 3(1): 1–105.

—, 1927b. New and interesting Foraminifera from Mexico and Texas. *Contrib. Cushman Lab. Foram. Res.*, 3(2): 111–119.

—, 1931. *Hastigerinella* and other interesting Foraminifera from the Upper Cretaceous of Texas. *Contrib. Cushman Lab. Foram. Res.*, 7(4): 83.

—, 1946. Upper Cretaceous Foraminifera of the Gulf Coastal region of the United States and adjacent areas. *U.S. Geol. Surv. Profess. Papers*, 206: 1–241.

—, 1951. Paleocene Foraminifera of the Gulf Coastal region of the United States and adjacent areas. *U.S. Geol. Surv. Profess. Papers*, 232: 1–75.

Cushman, J. A. and Bermudez, P. J., 1949. Some Cuban species of *Globorotalia*. *Contrib. Cushman Lab. Foram. Res.*, 25(2): 26–45.

Cushman, J. A. and Ten Dam, A., 1948. *Globigerinelloides*, a new genus of the Globigerinidae. *Contrib. Cushman Lab. Foram. Res.*, 24: 42–43.

Cushman, J. A. and Ellisor, A. C. 1939. New species of foraminifera from the Oligocene and Miocene. *Contrib. Cushman Lab. Foram. Res.*, 15(1): 1–14.

Cushman, J. A. and Jarvis, P. W., 1936. Three new Foraminifera from the Miocene Bowden marl of Jamaica. *Contrib. Cushman Lab. Foram. Res.*, 12(1): 3–5.

Cushman, J. A. and Renz, H. H., 1942. Eocene, Midway, Foraminifera from Soldado Rock, Trinidad. *Contrib. Cushman Lab. Foram. Res.*, 18(1): 1–20.

—, 1946. The foraminiferal fauna of the Lizard Springs formation of Trinidad, B.W.I. *Cushman Lab. Foram. Res., Spec. Publ.*, 18: 1–48.

—, 1948. Eocene Foraminifera of the Navet and Hospital Hill formation of Trinidad, B.W.I. *Cushman Lab. Foram. Res., Spec. Publ.*, 24: 1–42.

Cushman, J. A. and Stainforth, R. M., 1945. The Foraminifera of the Cipero marl formations of Trinidad, British West Indies. *Cushman Lab. Foram. Res., Spec. Publ.*, 14: 1–75.

Cushman, J. A. and Todd, R., 1948. A foraminiferal fauna from the New Almaden district, California. *Contrib. Cushman Lab. Foram. Res.*, 24(4): 90–98.

Cuvillier, J., Dalbiez, F., Glintzboeckel, C., Perebaskine, V., Lys, M., Magné, J. and Rey, M., 1955. Etudes micropaléontologiques de la limite Crétacé–Tertiaire dans les mers mésogéennes. *World Petrol. Congr., Proc., 4th, Rome, 1955*, Sect. I(D) 6: 517–544.

Dalbiez, F., 1955. The genus *Globotruncana* in Tunisia. *Micropaleontology*, 1(2): 161–171.

—, 1956. Etude sommaire des microfaunes de la région du Kef (Tunisie). *Carte géologique de la Tunisie*. Feuille 44.

De Klasz, J., 1953. Einige neue oder wenig bekannte Foraminiferen aus der helvetischen Oberkreide der Bayerischen Alpen südlich Traunstein (Oberbayern). *Geol. Bavarica*, 17: 223–244.

—, 1956. Zur Kenntnis der ostalpinen Oberkreidestratigraphie. *Neues Jahrb. Geol. Paläontol., Monatsh.*, 9: 410–419.

—, 1961. Présence de *Globotruncana concavata* (Brotzen) et *Gtr. concavata carinata* Dalbiez (Foraminifères) dans le Coniacien du Gabon (Afrique équatoriale). *Compt. Rend. Soc. Géol. France, 1961*, 5: 123–124.

De Lapparent, J., 1918. Etude lithologique des terrains crétacés de la région d'Hendaye. In: *Mémoires pour Servir à l'Explication de la Carte géologique détaillée de la France*. Minist. des Travaux publics, Paris, pp.1–153.

De Stefani, T., 1950. Su alcune manifestazioni di idrocarburi in provincia di Palermo e descrizione di Foraminiferi nuovi. *Plinia, Ital.*, 3(4): 9.

Drooger, C. W., 1956. Transatlantic correlation of the Oligo-Miocene by means of Foraminifera. *Micropaleontology*, 2(2): 183–192.

—, 1964a. Problems of mid-Tertiary stratigraphic interpretation. *Micropaleontology*, 10(3): 369–374.

—, 1964b. Zonation of the Miocene by means of planktonic Foraminifera; a review and some comments. *Symp. Neogene Congr., 3rd, Bern, 1964*, pp.40–50.

Drooger, C. W. and Batjes, D. A. J., 1959. Planktonic Foraminifera in the Oligocene and Miocene of the North Sea Basin. *Koninkl. Ned. Akad. Wetenschap., Proc., Ser. B*, 62(3): 172–186.

Dunnington, H. V., 1955. Close zonation of Upper Cretaceous globigerinal sediments by abundance of *Globotruncana* species groups. *Micropaleontology*, 1(3): 207–219.

Eames, F. E., Banner, F. T., Blow, W. H. and Clarke, W. J., 1962. *Fundamentals of Mid-Tertiary Stratigraphical Correlation*. Cambridge University Press, London, 151 pp.

—, 1964. New Zealand Mid-Tertiary Stratigraphical Correlation. *Nature*, 203 (4941): 180–181.

—, 1965. Dating of some beds in Panama and Trinidad. *J. Paleontol.*, 39(1): 162–163.

El-Naggar, Z. R., 1969. Correlation of the various planktonic foraminiferal zonations of the Paleocene. *Proc. Intern. Conf. Planktonic Microfossils, 1st*, 2: 202–223.

Ericson, D. B., Ewing, M. and Wollin, G., 1963. Pliocene–Pleistocene boundary in deep-sea sediments. *Science*, 139 (3556): 727–737.

Finlay, H. J., 1939a. New Zealand Foraminifera: Key Species in Stratigraphy—No. 2. *Trans. Roy. Soc. New Zealand*, 69(1): 89–128.

—, 1939b. New Zealand Foraminifera: Key Species in Stratigraphy—No. 3. *Trans. Roy. Soc. New Zealand*, 69(3): 309–329.

—, 1947. New Zealand Foraminifera: Key Species in Stratigraphy—No. 5. *New Zealand J. Sci. Technol.*, 28(5): 259–292.

Follador, U., 1967. Il Pliocene ed il Pleistocene dell'Italia centro–meridionale, versante Adriatico. Biostratigrafia. *Boll. Soc. Geol. Ital.*, 86: 565–584.

Gandolfi, R., 1942. Ricerche micropaleontologiche e stratigrafiche sulla Scaglia e sul Flysch Cretacici dei dintorni di Balerna (Canton Ticino). *Riv. Ital. Paleontol., A.*, 48(4): 1–160.

—, 1955. The genus *Globotruncana* in northeastern Colombia. *Bull. Am. Paleontol.*, 36(155): 1–118.

414

GANSS, O. and KNIPSCHEER, H. C., 1956. Die Maastricht–Eozaen-Folge des Helveticums im Sprunggraben bei Obereisendorf (Abb.) und ihre Gliederung mit Hilfe pelagischer Foraminiferen. *Geol. Jahrb.*, 71: 617–630.

GLAESSNER, M. F., 1937. Plankton Foraminiferen aus der Kreide und dem Eozän und ihre stratigraphische Bedeutung. In: *Studies in Micropaleontology*. Lab. Paleontol. Moscow Univ., Moscow, 1(1): 27–47.

GRIMSDALE, T. F., 1951. Correlation, age determination, and the Tertiary pelagic Foraminifera. *World Petrol. Congr., Proc., 3rd, The Hague, 1951*, 1: 464–475.

HAGN, H., 1956. Geologische und paläontologische Untersuchungen im Tertiär des Monte Brione und seiner Umgebung (Gardasee, Ober-Italien). *Paleontographica*, A 107: 67–210.

HAGN, H. and ZEIL, W., 1954. Globotruncanen aus dem Ober-Cenoman und Unter-Turon der Bayerischen Alpen. *Eclogae Geol. Helv.*, 47(1): 1–60.

HAMILTON, E. L., 1953. Upper Cretaceous, Tertiary and Recent planktonic Foraminifera from Mid-Pacific flat-topped seamounts. *J. Paleontol.*, 27(2): 204–237.

HAY, W. W., 1960. The Cretaceous–Tertiary Boundary in the Tampico Embayment, Mexico. *Intern. Geol. Congr., 21st, Copenhagen, 1960, Sect. 5*, pp.70–77.

HEDBERG, H. D., 1937. Foraminifera of the middle Tertiary Carapita formation of northeastern Venezuela. *J. Paleontol.*, 11(8): 661–697.

HERB, R., 1962. Geologie von Amden. *Beitr. Geol. Karte Schweiz*, 114: 1–130.

HERMES, J. J., 1960. A simplified method for cleaning Foraminifera. *Rev. Micropaléontol.*, 3(3): 155–156.

—, 1966. Lower Cretaceous planktonic Foraminifera from the Subbetic of southern Spain. *Geol. Mijnbouw*, 45: 157–164.

HORNIBROOK, N. DE B., 1958. New Zealand Upper Cretaceous and Tertiary foraminiferal zones and some overseas correlations. *Micropaleontology*, 4(1): 25–38.

—, 1967. New Zealand Tertiary microfossil zonation, correlation and climate. *Proc. Pacific Sci. Congr. Pacific Sci. Assoc., 11th, Tokyo, 1967*, pp.29–39.

HOTTINGER, L. and SCHAUB, H., 1960. Zur Stufeneinteilung des Paläocaens und Eocaens: Einführung der Stufen Ilerdien und Biarritzien. *Eclogae Geol. Helv.*, 53(1): 453–479.

HOWE, H. V. and WALLACE, W. E., 1932. Foraminifera of the Jackson Eocene at Danville landing on the Ouachita, Catahoula Parish, Louisiana. *Louisiana Dept. Conserv., Geol. Bull.*, 2: 7–118.

JENKINS, D. G., 1964a. Location of the Pliocene–Pleistocene boundary. *Contrib. Cushman Found. Foram. Res.*, 15(1): 25–27.

—, 1964b. Panama and Trinidad Oligocene Rocks. *J. Paleontol.*, 38(3): 606.

—, 1964c. New Zealand Mid-Tertiary Stratigraphical Correlation. *Nature*, 203 (4941): 181–182.

—, 1965. The origin of the species *Globigerinoides trilobus* (REUSS) in New Zealand. *Contrib. Cushman Found. Foram. Res.*, 16(3): 116–120.

—, 1967. Planktonic foraminiferal zones and new taxa from the Lower Miocene to the Pleistocene of New Zealand. *New Zealand J. Geol. Geophys.*, 10(4): 1064–1078.

KNIPSCHEER, H. C. G., 1956. Biostratigraphie in der Oberkreide mit Hilfe der Globotruncanen. *Paläontol. Z.*, 30: 50–56.

KOCH, R., 1926. Mitteltertiäre Foraminiferen aus Bulongan, Ost-Borneo. *Eclogae Geol. Helv.*, 19(3): 722–751.

—, 1935. Namensänderung einiger Tertiär-Foraminiferen aus Niederländisch Ost-Indien. *Eclogae Geol. Helv.*, 28(2): 557–558.

LAGAAIJ, R., 1963. *Cupuladria canariensis* (BUSK)—portrait of a Bryozoan. *Palaeontology*, 6(1): 172–217.

LALICKER, C. G., 1948. A new genus of Foraminifera from the Upper Cretaceous. *J. Paleontol.*, 22(5): 624.

LE ROY, L. W., 1939. Some small Foraminifera, ostracods and otoliths from the Neogene ("Miocene") of the Rokan–Tapanoeli area, Central Sumatra. *Natuurk. Tijdschr. Ned.-Indië*, 99(6): 215–296.

—, 1952. *Orbulina universa* D'ORBIGNY in Central Sumatra. *J. Paleontol.*, 26(4): 576–584.

—, 1953. Biostratigraphy of the Maqfi section, Egypt. *Geol. Soc. Am., Mem.*, 54: 1–73.

LIPPS, J. H., 1967. Planktonic Foraminifera, intercontinental correlation and age of California Mid-Cenozoic microfaunal stages. *J. Paleontol.*, 41(4): 994–999.

LOEBLICH, A. R. and TAPPAN, H., 1946. New Washita Foraminifera. *J. Paleontol.*, 20(3): 238–258.

—, 1957a. The new planktonic foraminiferal genus *Tinophodella* and an emendation of *Globigerinita* BRÖNNIMANN. *J. Wash. Acad. Sci.*, 47(4): 112–116.

—, 1957b. Planktonic Foraminifera of Paleocene and early Eocene age from the Gulf and Atlantic Coastal Plains. *U.S. Natl. Museum, Bull.*, 215: 173–198.

—, 1957c. Correlation of the Gulf and Atlantic Coastal Plain Paleocene and Lower Eocene formations by means of planktonic Foraminifera. *J. Paleontol.*, 31(6): 1109–1137.

—, 1961. Cretaceous planktonic Foraminifera: Part 1—Cenomanian. *Micropaleontology*, 7(3): 257–304.

LOEBLICH, A. R. et al., 1957. Studies in Foraminifera. *U.S. Natl. Museum, Bull.*, 215: 1–323.

LUTERBACHER, H., 1964. Studies in some *Globorotalia* from the Paleocene and lower Eocene of the Central Apennines. *Eclogae Geol. Helv.*, 57(2): 631–730.

LUTERBACHER, H. and PREMOLI SILVA, I., 1964. Biostratigrafia del limite Cretaceo–Terziario nell'Appennino Centrale. *Riv. Ital. Paleontol. Stratigraf.*, 70(1): 67–128.

MALAPRIS, M. and RAT, P., 1961. Données sur les Rosalines du Cénomanien et du Turonien de Côte-d'Or. *Rev. Micropaläontol.*, 4(2): 85–98.

MARIE, P., 1941. Les foraminifères de la craie à *Belemnitella mucronata* du bassin de Paris. *Mém. Museum Natl. Hist. Nat., Paris, N. Ser.*, 12: 1–296.

—, 1947. Sur quelques *Rosalinella* du sondage de Peyrehorade. *Compt. Rend. Soc. Géol. France*, 16 1961: 124–125.

—, 1961. Présence de *Globotruncana elevata* (BROTZEN) dans le Santonien des Corbières. *Compt. Rend. Soc. Geol. France*, 5.

MARTIN, L. T., 1943. Eocene Foraminifera from the type Lodo Formation, Fresno County, California. *Stanford Univ. Publ., Univ. Ser., Geol. Sci.*, 3(3): 93–125.

MISTRETTA, F., 1962. Foraminiferi planctonici del Pliocene inferiore di Altavilla Milicia (Palermo, Sicilia). *Riv. Ital. Paleontol.*, 68(1): 97–114.

MORNOD, L., 1949. Les Globorotalidés du Crétacé supérieur du Montsalvens (Préalpes fribourgeoises). *Eclogae Geol. Helv.*, 42(2): 573–596.

MOROZOWA, V. G., 1957. The foraminiferal superfamily Globigerinidea superfam. nova and some of its representatives. *Dokl. Akad. Nauk S.S.S.R.*, 114(5): 1109–1112.

MORROW, A. Z., 1934. Foraminifera and Ostracoda from the Upper Cretaceous of Kansas. *J. Paleontol.*, 8(2): 186–205.

NAKKADY, S. E., 1950. A new foraminiferal fauna from the Esna shales and Upper Cretaceous Chalk of Egypt. *J. Paleontol.*, 24(6): 675–692.

—, 1951. Zoning the Mesozoic–Cenozoic formation of Egypt by the Globorotaliidae. *Bull. Fac. Sci. Farouk I Univ.*, 1: 45–58.

NUTTALL, W. L. F., 1928. Notes on the Tertiary Foraminifera of southern Mexico. *J. Paleontol.*, 2(4): 372–376.

—, 1930. Eocene Foraminifera from Mexico. *J. Paleontol.*, 4(3): 271–293.

—, 1932. Lower Oligocene Foraminifera from Mexico. *J. Paleontol.*, 6(1): 3–35.

OLSSON, R. K., 1964. *Praeorbulina* OLSSON, a new foraminiferal genus. *J. Paleontol.*, 38(4): 770–771.

ONOFRIO, S. D', 1964. I foraminiferi del neostratotipo del Messiniano. *Giorn. Geol., Ann. Museo Geol. Bologna* 32(2): 409–461.

PALMER, D. K., 1945. Notes on the Foraminifera from Bowden, Jamaica. *Bull. Am. Paleontol.*, 29(115): 1–82.

PARKER, F. L., 1962. Planktonic foraminiferal species in Pacific sediments. *Micropaleontology*, 8(2): 219–254.

—, 1965. A new planktonic species (Foraminiferida) from the Pliocene of Pacific deep-sea cores. *Contrib. Cushman Found. Foram. Res.*, 16(4): 151–152.

PERCONIG, E., 1968. Bioestratigrafía del Neogeno Mediterraneo basada en los Foraminiferos planktonicos. *Rev. Esp. Micropaleontol.*, 1(1): 103–111.

PESSAGNO, E. A., 1967. Upper Cretaceous planktonic Foraminifera from the Western Gulf Coastal Plain. *Palaeontograph. Am.*, 5(37): 243–545.

PETTERS, V., 1954. Tertiary and Upper Cretaceous Foraminifera from Colombia, S.A. *Contrib. Cushman Found. Foram. Res.*, 5(1): 37–41.

PLUMMER, H. J., 1926. Foraminifera of the Midway formation in Texas. *Univ. Texas Bull.*, 2644: 1–206.

—, 1931. Some Cretaceous Foraminifera in Texas. *Univ. Texas Bull.*, 3101: 109–203.

POAG, C. W. and AKERS, W. H., 1967. *Globigerina nepenthes* TODD of Pliocene age from the Gulf Coast. *Contrib. Cushman Found. Foram. Res.*, 18(4): 168–176.

PREMOLI SILVA, I. and PALMIERI, V., 1962. Osservazioni stratigrafiche sul Paleogene della Val di Non (Trento). *Mem. Soc. Geol. Ital.*, 3: 191–212.

RAMSAY, W. R., 1962. Hantkenininae in the Tertiary rocks of Tanganyika. *Contrib. Cushman Found. Foram. Res.*, 13(3): 79–89.

REICHEL, M., 1948. Les Hantkéninidés de la Scaglia et des Couches rouges (Crétacé supérieur). *Eclogae Geol. Helv.*, 40(2): 391–409.

—, 1949. Observations sur les *Globotruncana* du gisement de la Breggia (Tessin). *Eclogae Geol. Helv.*, 42(2): 596–617.

—, 1952. Remarques sur les Globigérines du Danien de Faxe (Danemark) et sur celles des couches de passage du Crétacé au Tertiaire dans la Scaglia de l'Apennin. *Eclogae Geol. Helv.*, 45(2): 341–349.

REISS, Z., 1952. On the Upper Cretaceous and Lower Tertiary microfaunas of Israel. *Bull. Res. Council. Israel, Sect. B*, 2(1): 37–49.

—, 1955. Micropaleontology and the Cretaceous–Tertiary boundary in Israel. *Bull. Res. Council Israel, Sect. B*, 5(1): 105–120.

—, 1957. The Bilamellidae nov. superfam. and remarks on Cretaceous Globorotaliids. *Contrib. Cushman Found. Foram. Res.*, 8(4): 127–145.

—, 1958. Classification of lamellar Foraminifera. *Micropaleontology*, 4(1): 51–70.

—, 1968. Planktonic foraminiferids, stratotypes and a reappraisal of Neogene chronostratigraphy in Israel. *Israel J. Earth-Sci.*, 17: 153–169.

416

RENZ, O., 1936. Stratigraphische und Mikropalaeontologische Untersuchung der Scaglia (Obere Kreide–Tertiär) im Zentralen Apennin. *Eclogae Geol. Helv.*, 29(1): 1–149.

RIEDEL, W. R., BRAMLETTE, M. N. and PARKER, F. L., 1963. Pliocene–Pleistocene boundary in deep-sea sediments. *Science*, 140 (3572): 1238–1240.

SAITO, T., 1962. Eocene planktonic Foraminifera from Hahajima (Hillsborough Island). *Paleontol. Soc. Japan, Trans. Proc.*, 45: 209–225.

SAITO, T. and BE, A. W. H., 1964. Planktonic Foraminifera from the American Oligocene. *Science*, 145 (3633): 702–705.

SANDIDGE, I. R., 1932. Foraminifera from the Ripley Formation of western Alabama. *J. Paleontol.*, 6(3): 265–287.

SCHACKO, G., 1896. Beitrag über Foraminiferen aus der Cenoman-Kreide von Moltzow in Mecklenburg. *Arch. Ver. Freunde Naturges. Mecklenburg*, 50.

SCHIJFSMA, E., 1955. La position stratigraphique de *Globotruncana helvetica* BOLLI en Tunisie. *Micropaleontology*, 1(4): 321–334.

SIGAL, J., 1948a. Notes sur les genres de Foraminifères *Rotalipora* BROTZEN, 1942 et *Thalmanninella*, famille des Globorotaliidae. *Rev. Inst. Franç. Pétrole Ann. Combust. Liquides*, 3(4): 95–103.

—, 1948b. Précisions sur quelques foraminifères de la famille des Globorotaliidae. *Compt. Rend. Soc. Géol. France*, 2.

—, 1952. Aperçu stratigraphique sur la micropaléontologie du Crétacé. *Congr. Géol. Intern., Compt. Rend., 19e, Algiers, 1952*, 26: 1–47.

—, 1955. Notes micropaléontologiques nord-africaines, 1. Du Cénomanien au Santonien: zones et limites en faciès pélagique. *Compt. Rend. Soc. Géol. France*, 8: 157–160.

—, 1958. La classification actuelle des familles de Foraminifères planktoniques du Crétacé. *Compt. Rend. Soc. Géol. France*, 12: 262–265.

SPRAUL, G. L., 1963. Current status of the Upper Eocene foraminiferal guide fossil *Cribrohantkenina*. *J. Paleontol.*, 37(2): 366–370.

STAINFORTH, R. M., 1960. Current status of Trans-Atlantic Oligo–Miocene correlation by means of planktonic foraminifera. *Rev. Micropaléontol.*, 2(4): 219–230.

—, 1964. Subdivision of the Miocene. *Bull. Am. Assoc. Petrol. Geologists*, 48(11): 1847–1848.

SUBBOTINA, N. N., 1953a. Fossil Foraminifera of the U.S.S.R. *Tr. Vses. Neft. Nauchn.-Issled Geolog. rasved. Inst.*, 76: 1–294.

—, 1953b. *Foraminifères fossiles d'U.R.S.S., Globigerinidae, Globorotaliidae, Hantkeninidae.* (Traduction M. Sigal.) Bur. Rech. Géol. et Min., Service d'inf. Géol., Traduction No. 2239, 306 pp.

SZÖTS, E., 1962. Remarques sur le problème de l'Oligocène américain et sur les zones planctoniques de l'Oligocène et du Miocène inférieur. *Compt. Rend. Soc. Géol. France*, 8: 236–237.

SZÖTS, E., MALMOUSTIER, G. and MAGNÉ, J., 1964. Observations sur le passage Oligocène–Miocène en Aquitaine et sur les zones de Foraminifères planctoniques de l'Oligocène. *Mém. Bur. Rech. Géol. Min., Colloq. Paléogène, 28, Bordeaux, 1962*, 1: 433–454.

TAKAYANAGI, Y. and SAITO, T., 1962. Planktonic Foraminifera from the Nobori Formation, Shikotu, Japan. *Sci. Rept. Tohoku Univ., Ser. 2, Spec. Papers*, 5: 67–105.

TAPPAN, H., 1943. Foraminifera from the Duck Creek formation of Oklahoma and Texas. *J. Paleontol.*, 17(5): 476–514.

THALMANN, H. E., 1932. Die Foraminiferen-Gattung *Hantkenina* CUSHMAN, 1924 und ihre regional-stratigraphische Verbreitung. *Eclogae geol. Helv.*, 25(2): 287–292.

—, 1934–1935. Die regional-stratigraphische Verbreitung der oberkretazischen Foraminiferen-Gattung *Globotruncana* CUSHMAN, 1927. Weitere Vorkommen von *Globotruncana* in der Oberkreide. *Eclogae Geol. Helv.*, 27(2): 413–428.

—, 1942a. *Hantkenina* in the Eocene of East Borneo. *Stanford Univ. Publ., Geol. Sci.*, 3(1).

—, 1942b. Foraminiferal genus *Hantkenina* and its subgenera. *Am. J. Sci.*, 240: 809–820.

—, 1959. New names for foraminiferal homonyms. IV. *Contrib. Cushman Found. Foram. Res.*, 10(4): 130–131.

TILEV, N., 1951. Etude des Rosalines Maestrichtiennes (genre *Globotruncana*) du Sud-Est de la Turquie (Sondage de Ramandag). *Bull. Lab. Géol., Mineral., Géophys. Musée Géol. Univ. Lausanne*, 103: 1–101.

TROELSEN, J. C., 1957. Some planktonic Foraminifera of the type Danian and their stratigraphic importance. *U.S. Natl. Museum Bull.*, 215: 125–132.

VELLA, P., 1964. Correlation of New Zealand and European Middle Tertiary. *Bull. Am. Assoc. Petrol. Geologists*, 48(12): 1938–1941.

—, 1965. Oligocene–Miocene Boundary. *Geol. Soc. Am. Bull.*, 76(3): 349–356.

VLERK, I. M. v. d. and POSTUMA, J. A., 1967. Oligo–Miocene Lepidocyclinas and planktonic Foraminifera from East Java and Madura, Indonesia. *Koninkl. Ned. Akad. Wetenschap., Proc., Ser. B*, 70(4): 391–398.

VOGLER, J., 1941. Ober-Jura und Kreide von Misol (Niederländisch-Ostindien). *Palaeontographica*, 4(4): 243–293.

VON HILLEBRANDT, A., 1962. Das Paleozän und seine Foraminiferenfauna im Becken von Reichenhall und Salzburg. *Abhandl. Bayer. Akad. Wiss. Math. Naturw. Kl.*, 108: 1–82.

WADE, M., 1964. Application of lineage concept to biostratigraphic zoning based on planktonic Foraminifera. *Micropaleontology*, 10(3): 273–290.

WEINZIERL, L. L. and APPLIN, E. R., 1929. The Claiborne Formation on the Coastal Domes. *J. Paleontol.*, 3(4): 384–410.

WEISS, L., 1955. Planktonic index Foraminifera of northwestern Peru. *Micropaleontology*, 1(4): 301–318.

WHITE, M. P., 1928. Some index Foraminifera of the Tampico embayment area of Mexico. *J. Paleontol.*, 2(3): 177–215.

INDEX OF SPECIES